The Physics of Experiment Instrumentation Using MATLAB Apps

With Companion Media Pack

Other books in this series by the author

One Hundred Physics Visualizations Using MATLAB (2013)

More Physics with MATLAB (2015)

Cosmology with MATLAB (2016)

Beams and Accelerators with MATLAB (2018)

Stars and Space with MATLAB Apps (2020)

The Physics of Experiment Instrumentation Using MATLAB Apps

The Companion Media Pack is available online at
https://www.worldscientific.com/worldscibooks/10.1142/12165#t=suppl

1. Register an account/login at https://www.worldscientific.com
2. Go to: https://www.worldscientific.com/r/12165-suppl
3. Download the Media Pack from
   https://www.worldscientific.com/worldscibooks/10.1142/12165#t=suppl

   For subsequent download, simply log in with the same login details in order to access.

The Physics of Experiment Instrumentation Using MATLAB Apps

With Companion Media Pack

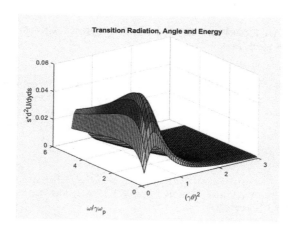

Transition Radiation, Angle and Energy

Dan Green

Fermi National Accelerator Laboratory, USA

 **World Scientific**

NEW JERSEY · LONDON · SINGAPORE · BEIJING · SHANGHAI · HONG KONG · TAIPEI · CHENNAI · TOKYO

Published by

World Scientific Publishing Co. Pte. Ltd.
5 Toh Tuck Link, Singapore 596224
USA office: 27 Warren Street, Suite 401-402, Hackensack, NJ 07601
UK office: 57 Shelton Street, Covent Garden, London WC2H 9HE

British Library Cataloguing-in-Publication Data
A catalogue record for this book is available from the British Library.

THE PHYSICS OF EXPERIMENT INSTRUMENTATION USING MATLAB APPS
With Companion Media Pack

ISBN 978-981-123-243-5 (hardcover)
ISBN 978-981-123-383-8 (paperback)
ISBN 978-981-123-244-2 (ebook for institutions)
ISBN 978-981-123-245-9 (ebook for individuals)

For any available supplementary material, please visit
https://www.worldscientific.com/worldscibooks/10.1142/12165#t=suppl

Preface

"We especially need imagination in science. It is not all mathematics, nor all logic, but it is somewhat beauty and poetry."

— **Maria Mitchell**

"Science is beautiful when it makes simple explanations of phenomena or connections between different observations. Examples include the double helix in biology and the fundamental equations of physics."

— **Stephen Hawking**

Some 20 years ago the author published a book on *The Physics of Particle Detectors*. It is timely to bring such a text up to date. Detectors have evolved greatly, enabled by the explosive progress of the microelectronics industry. Detectors with millions of elements operating at nanosecond even picosecond, time scales are now common. Because accelerators are the enabling technology of high-energy physics, new accelerators have enabled discoveries in particle physics. For example the CERN proton–antiproton storage ring enabled the discovery of the W and Z bosons in 1983. The advent of the Fermilab Tevatron collider enabled the discovery of the top quark in 1995. Most recently the completion of the LHC at CERN led to the discovery of the Higgs boson in 2012. For this reason, the text describes particle detector, particle beam and accelerator instrumentation in a unified and coherent fashion.

In order to describe the instrumentation the underlying physics is unitary but the tools available to explore that physics have also improved enormously. The basic interactions are electromagnetic operating between photons and charged particles. Atomic systems are paramount; nuclei are normally inert.

There have been, over the last 20 years, many advances in Information Technology (IT). There are tools to enhance the learning process by means of visualization of the changes to a system operation in real time as the parameters of that system are altered. In addition, the advent of powerful internet search engines and the enormous expansion of available scientific material on the web makes more classical references appear slow and less accessible. A long list of references has not been created for this text.

Together, the hands on software applications and the enhanced accessibility of additional reference material using search engines makes for a vastly improved learning environment. This text uses Matlab Apps to connect a system to the fashion by which the system morphs under change in a single script and in real time. This strong connection makes the material more immediate, intuitive and comprehensible to the user. The Apps are available to the reader and can be run, read and modified as desired. They are a crucial and intrinsic component of the exposition and are meant to be explored concurrently with the text itself.

Additionally, liberal use of the figures and graphics available on the internet is made, so as to take advantage of the resources that can be found there. Many figures are generated by the Apps, but equally many are also downloaded from search engines. Indeed, preprint, open access and archives now make current results of academic research readily and widely available and very timely because they are often accessible much earlier before the material is actually published.

Introduction

"My religion consists of a humble admiration of the illimitable superior spirit who reveals himself in the slight details we are able to perceive with our frail and feeble mind."

— **Albert Einstein**

"Atoms are very special: they like certain particular partners, certain particular directions, and so on. It is the job of physics to analyze why each one wants what it wants."

— **Richard P. Feynman**

The goal of this text is to explore the physics of experimental instrumentation, in detectors, in beam lines and in accelerators in a unified and coherent manner. Rather than presenting the physics in a single formal exposition of all the topics in one place, they first appear in a brief summary at the start of each section. Following that, they are introduced only when necessary to understand a specific topic. The idea was to introduce the physics as it was needed to understand the instrument thus directly connecting physics concepts to their implementation. This approach is designed to make immediate connections between devices and the physics needed to understand their operation.

The text is divided into four main chapters. The first chapter, "Getting Started with Matlab", provides a basic introduction to the main tool used in this text, Matlab. Matlab is a vector language, well suited to describe experiment instrumentation. Matlab is available at low cost to the student. In addition, many colleges and businesses have site-wide Matlab licenses. Tools to convert from Matlab to

Python and C++ are also available so that the user can change the environment as desired. Finally, Matlab is not absolutely necessary since each App is illustrated with a figure which shows the results of a particular set of the parameters chosen by the user.

The second chapter, "Basic Physical Processes", introduces some basic notation and properties of atoms and nuclei. This text uses exclusively MKS/SI units in order to give a unified exposition. Motion in constant electric and magnetic fields is introduced in this section. Because these fields are the primary means of creating and controlling beams, these topics are covered early on and referred to later as needed. A table of all the symbols as well as the acronyms is given in the appendices at the end of the text.

The third chapter is called "Detector and Beam Instrumentation". Specific physics topics relevant to both beams and particle detector instrumentation are discussed when they are needed. In this way the physics is seen immediately prior to its first application. Beam devices used to measure position, momentum and particle type are explored. The idea here is to immediately tie the physics to the device.

The fourth chapter is titled "Accelerator Instrumentation". As was the case for chapter 3, some introductory accelerator specific physics topics are presented first. The associated devices are then studied with physics topics introduced only when needed to explore a specific device. The use of Matlab in both Chapters 3 and 4 particularly exploits the vector nature of the language and the symbolic logic tools in order to avoid lengthy and tedious calculations which are better left to a laptop. The goal is to help the user avoid what is avoidable; calculus, differential equations, transforms and the like. The focus is on the physics.

Experiment instrumentation is discussed in this book in a unified fashion for particle detectors, beam lines and accelerators. In many cases the same physical principles are used. Indeed, many of the techniques used are in common. Since accelerators are the enabling technology for high-energy physics it makes good sense to treat the instrumentation of the three specialties in unified fashion. Because the evolution of accelerator in high-energy physics has recently been toward colliders, there is an increasingly intimate connection between experiments and accelerators. That connection culminates in the interaction regions where the accelerator low beta insertion meets the detectors of the high-energy physicist. Clearly, a good understanding of all the composite technologies is of crucial importance to insure success of the experiments.

The topics covered in this text are severely limited to the physics of some of the common devices used in beam lines, experiments and accelerators. Discussion beyond the formation of a signal in a device is not covered. That limitation means that very interesting topics like triggering, Field Programmable Gate Arrays (FPGA), fiber optic data transmission, machine learning, high speed computing and a host of other topics are absent. The reader is encouraged to go beyond the limitations of this text to explore a much wider range of topics.

Contents

Chapter 1

Getting Started with Matlab

"The study of mathematics, like the Nile, begins in minuteness but ends in magnificence."

— **Charles Caleb Colton**

"Access to computers and the Internet has become a basic need for education in our society."

— **Kent Conrad**

1.1. Command Window

The Matlab product is available for users at many universities and businesses via site-wide licenses. When Matlab is first executed the Command window is opened. The ≫ prompt indicates Matlab is ready for an input command. Figure 1.1 shows the "Command" window just after invoking Matlab. Added displays of the command history and the workspace variables are useful and can be docked at the periphery of the Command Window using the "Home" tab with "Layout" selected in the dropdown menu.

Using the "Search Documentation" window on the top right, first search for "getting started". This opens the Matlab Documentation which has many choices and tutorials. Detailed responses to queries can also be obtained using the "Search Documentation" window at any time. Shorter explanations are available via the Command window using "help sin" for example for the sin function. However, the name of the function must first be known.

The "Search Documentation" tab allows the user to search the extensive set of Matlab topics and tutorials. For example, a search

Figure 1.1: "Command" window appearance at the start of Matlab. The "Workspace" window and the "Command History" window have previously been added to the layout and docked.

on "function" opens up many explanations of the available functions. A search on "symbolic math" enables an explanation of the available functions and solvers and supplies an extensive tutorial for symbolic math tools using "Live" script.

The "Home" tab has an "Open" tab to open editors or create new scripts. The scripts supplied by this text can then be invoked and run. One first has to use the "set path" button to indicate where your scripts can be found or where user data is stored.

In this fashion, the new Matlab user can quickly search for help and quickly build up their expertise with the very extensive suite of Matlab capabilities. In future examples and Apps, new tools that are used in the scripts supplied in this text will be explained as they arise, case by case rather than in a large, indigestible, mass of documentation.

The "HOME" tab expands for the user as displayed in Figure 1.2. Note the "learn Matlab" tab. That tab allows access to several online courses that cover a wide variety of topics for early users. The major use of this tab in this text is to create new scripts and Apps, using

Figure 1.2: The drop down "HOME" tab in the "Command" window. New scripts can be created or existing scripts or Apps can be opened in the "Editor" window.

the "New" or "Open" tab and to use the "Editor" to write, save, open, and debug scripts in the "Editor" window. In all cases in this text scripts will be opened from the "Command" window. There is a separate editor for Apps which can be invoked with "New"/"App" or "Open"/"App" and then choosing from a list of existing Apps.

1.2. Editor

Using the "Open" tab the user will invoke the "Editor" window. For an existing script the tab lists scripts which are on the path which was set initially. Choosing "Editor_Script.m" opens the script in the "Editor" window as shown in Figure 1.3. The example shown below shows the color coding of the instructions. The main tab is the "Editor" with options which are largely self-explanatory.

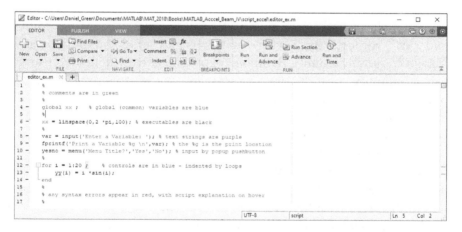

Figure 1.3: The result of choosing "Open" in the "Command Window" and then choosing the script "Editor_Script.m" which is one of the scripts provided for use with this text.

During script creation, using the "New" tab, the Editor indicates incorrect statement lines and what the issue might be — in red. Hovering on the name of a utility brings up a short indication of the required syntax, Comments, with a leading % character, are in green. Global variables are shown in teal. Printout in the "Command Window" is suppressed by ending a line in;. Input variables are entered in the "Command Window" by using "input" to ask the user for command line input. There is also a popup "menu" utility which allows the user to click on specified choices. Output variables are printed in the "Command Window" using the command "fprintf". Executable operations are shown in black, while text strings appear as purple characters. Clearly this is a lot of material to take in for a new user, so several simple examples are given below to ease into the first use of Matlab. The Matlab tutorials are also very useful for beginning users.

Matlab is a vector/matrix language, which makes it ideal for accelerator applications. In addition it has a full suite of symbolic math tools. One need never look up a differential, integral, special function, or algebraic and differential systems of equations. These are all available and allow one to avoid the drudgery while focusing on the physics.

Symbolic differentiation and integration are done using "diff" and "int", respectively. Taylor expansions and series summations use "taylor" and "symsum". Symbolic expressions can be made clearer using "simplify" and "pretty" or by using "subs" to substitute simplifications. Algebraic equations and differential equations are solved using "solve" and "dsolve". If a closed form solution does not exist "ode45" performs numerical evaluation of the solutions of differential equations. First-order partial differential equations can also be solved numerically. The mathematics for both beams and accelerators was developed in the 19th and early 20th centuries. Those tools are by now codified and readily available in Matlab.

Most special functions are available, for example Bessel functions. Matrix transpose and inverse are also available operations. Appendix A shows the Matlab scripts for Section 2 as an aid to see the coding of some simple scripts and Apps. In general, all the scripts used in the text are available to the user. It is intended that for a given topic the user combines the text, the execution of the App and some "playing" with the variables in real time so that the user gets some intuition as the how the parameters affect the solutions. Many of the scripts make "movies" where a display of the evolution of a system in time is produced for the user.

1.3. Online Resources

There are also several online courses offered by the makers of Matlab. A browser search for "Matlab" is a rich source of possibilities. An example appropriate for just getting started is shown in Figure 1.4. Another option is to utilize "Getting Started With Matlab".

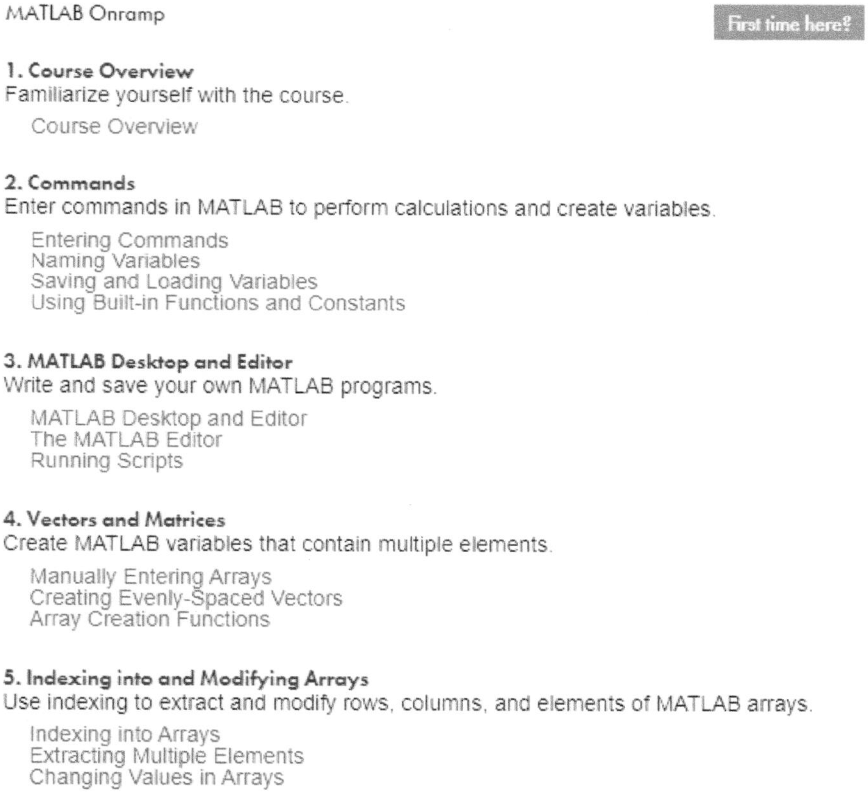

MATLAB Onramp

First time here?

1. Course Overview
Familiarize yourself with the course.

Course Overview

2. Commands
Enter commands in MATLAB to perform calculations and create variables.

Entering Commands
Naming Variables
Saving and Loading Variables
Using Built-in Functions and Constants

3. MATLAB Desktop and Editor
Write and save your own MATLAB programs.

MATLAB Desktop and Editor
The MATLAB Editor
Running Scripts

4. Vectors and Matrices
Create MATLAB variables that contain multiple elements.

Manually Entering Arrays
Creating Evenly-Spaced Vectors
Array Creation Functions

5. Indexing into and Modifying Arrays
Use indexing to extract and modify rows, columns, and elements of MATLAB arrays.

Indexing into Arrays
Extracting Multiple Elements
Changing Values in Arrays

Figure 1.4: Screen capture of a Matlab introductory course called "Matlab Onramp".

Chapter 2

Basic Physics Processes

"How far that little candle throws his beams. So shines a good deed in a weary world."

— **William Shakespeare**

"As far as we can discern, the sole purpose of human existence is to kindle a light in the darkness of mere being."

— **Carl Jung**

2.1. Physics Constants and Properties of Materials

As indicated in Eq. (2.1), atomic binding energy scales are a few electronvolts, implying atomic sizes of about 0.1 nm. Nuclear energy scales are a few mega electron volts, implying sizes of nuclei of about 1 fm, 100,000 times smaller. Planck's constant sets the fundamental quantum mechanical limit on the uncertainty of measurements of energy and time, position and momentum and all other non-commuting observables. It is expected that atomic energies are, very approximately, ~10 eV, while nuclear energies are ~10 MeV. The MKS unit for the cross-section of a physical process is the barn. All else being equal, the cross-section for an atomic process is much larger than that for a nuclear process, based on the sizes of the two systems.

$$0.1 \, \text{nm} = 10^{-10} \, \text{m}, \quad 1 \, \text{fm} = 10^{-15} \, \text{m}$$

$$\hbar c = 200 \, \text{eVnm} = 0.2 \, \text{GeVfm} \tag{2.1}$$

$$1b = 10^{-28} \, \text{m}^2, \quad (\hbar c)^2 = 0.4 \, \text{GeV}^2 \quad mb = 4 \times 10^{14} \, \text{eV}^2 b$$

The MKS units of $e^2/(4\pi\varepsilon_o)$ are energy times distance, so that α is dimensionless, as seen in Eq. (2.2). It sets the scale for the coupling strength of the electromagnetic field to matter. The reduced Compton wavelength of the electron is 0.000391 nm. The classical electron radius is that radius when its electromagnetic self-energy is approximately equal to its rest energy $m_e c^2$, or 2.82 fm. Only the reduced Compton wavelength will be used in this text in general, but the Compton wavelength will be mentioned in the later section on Compton scattering. As well, the reduced Planck constant will be used in the text in almost all cases.

$$\alpha = (e^2/4\pi\varepsilon_o)/\hbar c = 1/137$$

$$r_e = (\alpha\hbar c)/m_e c^2, \quad \lambdabar = \hbar/(m_e c) \tag{2.2}$$

$$r_e = \alpha\lambdabar$$

A few basic physics constants are shown in a table in Appendix B. In Appendix C, some basic properties of selected atoms are also displayed. Appendix D gives the index of refraction of several materials. In Appendix E, a table of the symbols used in this text is provided. More extensive values of parameters, when required, will be provided in the text. Additional data can be found using any reasonable browser in a search. Acronyms used in the text appear in Appendix F.

In the text itself, most calculations are sketched out explicitly, but almost never in full textbook rigor. The aim of the text is not to derive all the physics, but to make it plausible when it is applied to a particular system used to detect beams of particles or beams contained in accelerators. Beyond the generalities of this section, specific physics processes will be developed when they are needed to understand a given detector type.

2.2. Atoms

The simplest atomic system is hydrogen with one electron and one proton bound together by the electric attraction between them. The electron in hydrogen in the ground, or lowest energy, state is bound to the proton with an energy, ε_o, much less than the electron rest energy,

$-13.6\,\mathrm{eV}$ compared to $0.511\,\mathrm{MeV}$. That fact shows that electrons in atoms move in non-relativistic (NR) orbits. Less bound states have an energy that follows an inverse square series of integer quantum number, n, values with the first excited state, $n = 2$, bound by only $-3.4\,\mathrm{eV}$. The dependence of the binding energy on α^2 can be understood as due to the emission and absorption of a virtual photon between the electron and proton where each process "costs" a factor of α.

$$\varepsilon_o = -\alpha^2(m_e c^2)/2$$
$$\varepsilon_n = \varepsilon_o/n^2 \tag{2.3}$$

The radius of the most bound state is the first Bohr radius with a value of $0.053\,\mathrm{nm}$. The radii of other bound states are larger, growing as n increases because they are less tightly bound. The classical velocity with respect to c, β, is α which validates the NR properties of the hydrogen atom. In Eq. (2.4) there are two instances of the symbol ε_o, in one case it represents the hydrogen ground state energy, and in the other, the vacuum permittivity. This should not be confusing since the permittivity dependence is easily removed by using other constants.

$$a_o = \lambdabar/\alpha, \quad a_n = a_o n^2$$
$$\beta = \alpha \tag{2.4}$$
$$\varepsilon_o = -e^2/(8\pi\varepsilon_o a_o) = -\alpha(\hbar c/2a_o)$$

The radii of atoms in their ground state are all within an order of magnitude of the hydrogen first Bohr radius, and are typically $0.1\,\mathrm{nm}$. The most bound inner electrons have energies approximately $\varepsilon_o Z^2$, since they "see" all the protons, unscreened by the other electrons. The structure in radius as a function of atomic number, Z, is shown in Figure 2.1. It is caused by the shell structure of the states defined by the quantum solutions for the ground state energies. That structure defines the periodic structure of the elements going from metals to noble gases. The ionization potential, I, is the energy needed to free an outer, loosely bound, atomic electron and has typical values 5–$25\,\mathrm{eV}$ comparable to the hydrogen ground state energy.

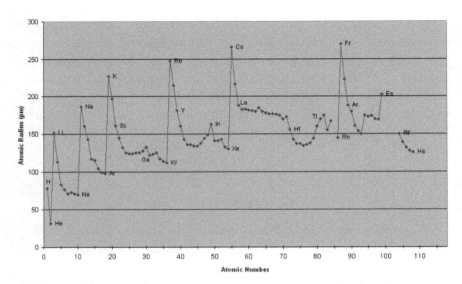

Figure 2.1:   Dependence of the atomic radius on the atomic number Z. The large radius elements are metals, while the small atomic radii are for the noble gases.

Metals have small values of I, insulators have large values, while semiconductors have intermediate values. The elements with loosely bound outer electrons are metals, while the elements with tightly bound electrons are the inert noble gases. The range of sizes is from about $0.03\,nm$ for helium to about $2.7\,nm$ for cesium. The ionization potential structure in Z is similar for each successive shell.

A basic concept is that of the cross-section. The cross-section is fundamental and defines the probability of a reaction to occur once the external properties of the density of targets and the beam flux are defined and factored out. It is defined completely by the physics and it has the dimensions of length squared. In this text, a cross-section will normally have the dimensions explicitly displayed with other dimensionless terms factored out. In the absence of any particular dynamics, the cross-section for the hydrogen atom is the geometric one defined simply by the size of the scattering object treated as a simple "billiard ball". The size of atomic cross-sections, given the similar physical sizes, will then be expected to be about 10^8 barns.

$$\sigma_H \sim \pi a_o^2 = \pi (\lambda/\alpha)^2 = 8.8 \times 10^7 \, b \qquad (2.5)$$

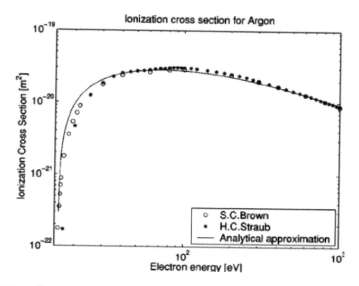

Figure 2.2: Cross-section for scattering with ionization for electron scattering off hydrogen. The threshold electron energy is the ground state binding energy.

The cross-section for the process of scattering off hydrogen which leads to ionizing the single electron is shown in Figure 2.2. The cross-section rises rapidly with incident electron energy as the binding energy of 13.6 eV of the ground state is exceeded. The peak value for the cross-section is about $3 \times 10^8 \, b$, near the value shown in Eq. (2.5). The metals may have cross-sections ten times larger based on their larger radii.

The geometric description for the cross-section only holds if the processes entail objects which are simple, hard spherical, geometric entities. That condition might apply in a dilute gas of atoms. If there is some dynamics, the cross-sections will differ as will be explored starting now and carrying throughout the balance of the chapter.

Begin with the Thomson cross-section for the scattering of low energy photons off a localized charge, taken here to be a single-free electron. The angular distribution is dipole. The cross-section for this process is smaller than the geometric cross-section by a factor α^2 because it occurs by way of the interaction of the photon with the electron via the electromagnetic field. At very low energies, the cross-section for an atom is just Z times the free electron cross-section because the photon scatters incoherently off all the electrons in the atom. The interaction is assumed to be elastic; the photon loses no energy in the process.

The cross-section is normalized to the photon energy flux using the Poynting vector, S, which is the cross product of the electric, E, and magnetic fields, B, of the photon and has dimensions of power, P, per unit area. For an incident plane wave with amplitude E_o of the electric field, the time-averaged Poynting vector is $\varepsilon_o c E_o^2 / 2$. The incident flux is not intrinsic to the physics of the process and so it is factored out leaving the cross-section, σ_T.

$$d\sigma_T/d\Omega = [d\langle P\rangle/d\Omega]/\langle|\vec{S}|\rangle, \vec{S} = \vec{E} \times \vec{B}$$

$$= (\alpha \lambda_e)^2 \sin^2 \theta \qquad (2.6)$$

$$\sigma_T = (8\pi/3)r_e^2 = (8\pi/3)(\alpha \lambda_e)^2 = 0.66\,b$$

The cross-section has units of length squared and is given in Eq. (2.6). In this case, the angular distribution is given as the cross-section per unit solid angle, where $d\Omega = d\phi \sin \theta d\theta$. In this text, it will be assumed that the processes are azimuthally symmetric unless explicitly stated otherwise so that $d\Omega = 2\pi \sin \theta d\theta$. Spherical polar coordinates will be used consistently, except in cases where Cartesian or cylindrical symmetry dictates using the appropriate coordinate systems.

The Thomson scattering is smaller than the geometric cross-section of πa_o^2 by a factor of about α^4. The scattering process can be thought of as the photon accelerating the electron, which then radiates at the same frequency, so that the photon energy is unchanged as if the photon were simply elastically scattered without change of energy. The cross-section to scatter off an atom, ignoring binding energy effects, would be expected to be Z times the Thomson cross-section since the scattering is incoherent and the scattering amplitudes do not interfere. The cross-section for photon scattering on copper is plotted later in the section on the photoelectric cross-section. Thomson scattering is important for low photon energies, while other processes dominate at higher energies.

Thinking of electron–photon scattering in quantum field theoretic terms, the process proceeds via the emission of a virtual electron emitted by the incoming electron and photon. That virtual electron then decays into a real electron and photon in the final state. The emission and decay both occur at an electromagnetic vertex, which implies a quantum amplitude proportional to e^2 or a reaction rate proportional to α^2. That factor is evident in Eq. (2.6).

2.3. Nuclei

The charged protons in an atom bind the Z electrons to the small nucleus. The typical size of a nucleus follows if it is assumed that the protons and neutrons are hard spheres with a radius approximately given by the Compton radius of a proton, which is ~ 1 fm. The volume of a nucleus containing A protons plus neutrons then scales as the third power of A. The geometric cross-section scales as the 2/3 power of A and the mean free path, Λ, between collisions as the 1/3 power of A, $\Lambda^{-1} = N_A \rho \sigma / A$ where N_A is Avogadro's number. The number of nuclei per unit volume is $N_A \rho / A$. The transverse density of the targets is factored out in defining the cross-section keeping only the physics of the process, as it was for the Thomson scattering cross-section.

For a single proton, the estimated geometric cross-section for a radius of 1 fm is about 30 mb, about 10^{10} times smaller than the geometric hydrogen cross-section since the nucleus is about 10^5 times smaller than the atom. The mean free path is often quoted with the density effect factored out, in ρz or kg/m^2 units. The mean free path may be defined in the literature in length units or in length times density units. In those latter units, the density is factored out leaving just fundamental quantities as was the case for the cross-section.

$$V_N = (4\pi/3)a_N^3, \quad a_N \sim \lambdabar_p A^{1/3}, \quad \lambdabar_p \sim 1\, fm,$$
$$\sigma \sim \pi \lambdabar_p^2 A^{2/3}, \quad \Lambda \rho \sim A^{1/3}/(\pi \lambdabar_p^2 N_A) \tag{2.7}$$

Avogadro's number can easily be understood since the mass density ρ is equal to $m_A(N/V) = n m_A$, where n is the number density. Avogadro's number is defined to be: $(N/V) = N_A(\rho/A)$, but

$(N/V) = \rho/Am_p$ so that $N_A \sim 1/m_p = 5.98 \times 10^{26}\,\mathrm{kg}^{-1}$ in the MKS units used to define mass and density.

$$\#nuclei/volume = (N_A\rho)/A$$
$$\Lambda^{-1} = \sigma(N_A\rho)/A, (\Lambda\rho)^{-1} = \sigma(N_A)/A \tag{2.8}$$

The plot shown in Figure 2.3 is created by the script "nuc_lm_A" and uses the nuclear interaction lengths given in Appendix C. It displays the expected $A^{1/3}$ behavior. An approximate value of the mean free path is $\Lambda\rho = 350\,(\mathrm{kg/m^2})A^{1/3}$ summed over all inelastic nuclear reactions. That function is plotted as the red line in the figure, which illustrates that the approximate dependence works well. The collision length includes elastic scattering as a collision so that it is shorter than the interaction length. The interaction is assumed to be a strongly interacting particle, such as a proton, interacting with a nucleus. The treatment of nuclei as black, absorbing spheres follows from the fact that the strong nuclear forces are of short range. This

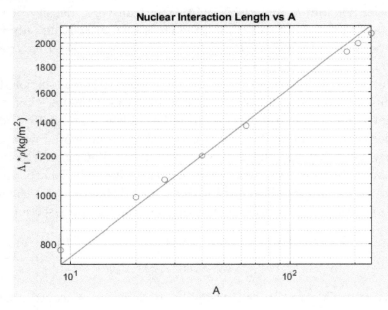

Figure 2.3: Plot of the inelastic nuclear mean free path times density, or interaction length, as a function of A. The red line has an $A^{1/3}$ dependence.

implies that the interactions occur only over a typical range of the strong force, which is approximately 1 fm.

A specific resource for data on properties of materials is the website of the "Particle Data Group" or PDG. The main goal of the PDG is to summarize all the accumulated knowledge relevant to fundamental particle physics. The front page of the PDG site is shown in Figure 2.4. This site has an enormous amount of data

Figure 2.4:   Front page of the website for the Particle Data Group.

on properties of materials. It also contains a specific section on particle detectors which complements well the material provided in the next two sections of this text. The reader is advised to look over and sample the materials provided at this site. However, one small caveat, MKS units are not always used, so conversions may be necessary.

2.4. Motion in Electric and Magnetic Fields

The instrumentation for detectors, beams and accelerators use electric and magnetic fields to control the charged particles which constitute, in aggregate, the beams and accelerated bunches of particles. The basic interactions are examined at this time in the simplest cases — uniform electric and magnetic fields. The first example is that of a uniform electric field, as approximately occurs in the interior of a parallel plate capacitor. The particle is assumed to have a quantized charge in q units of the magnitude of the electron charge $|e|$.

In both the non-relativistic (NR) and special relativistic (SR) cases, the momentum, p, increases linearly with time since qeE is the time rate change of momentum — the NR force. The electric field is E and is taken to be constant along the z direction. The NR acceleration, a, is a constant. The SR velocity approaches c asymptotically, but the momentum indeed increases in SR as $\beta cm\gamma$ not βcm, where $\gamma = 1/\sqrt{1-\beta^2}$. The NR limit occurs if at $\ll 1$, while for at $\gg 1$ the SR effects are very evident. A useful general check is to take the NR limit of an SR expression and establish that the classical result is obtained. More SR details are shown in Section 2.7. The equation of motion and its solution appear in Eq. (2.9).

The detailed motion is explored using the App "UniformE_App". Output is shown in Figure 2.5. A "movie" is made of the velocity and position in equal time intervals in the SR and NR cases. The movies are conceived to be animation tools where the equal time "frames" give a feel for the instantaneous velocity of the particle. A short description is given in the "Text" box. Symbolic expressions, found using "dsolve", for the momentum, velocity and position are

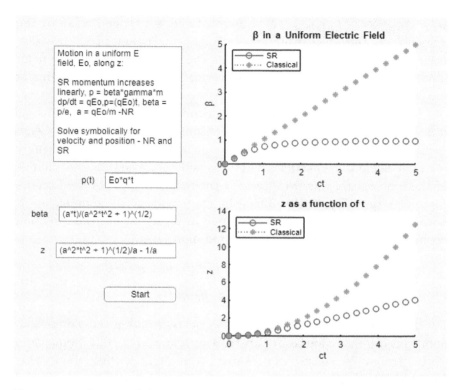

Figure 2.5: Output of the "UniformE_App". The solutions for SR and NR are both displayed for velocity and position as a function of time in the 2 plots. Both plots are displayed as "movies" in time frames.

displayed in the "EditFields". The button starts the App script.

$$dp/dt = qeE, \quad p = (qeE)t$$
$$\beta = (p/mc)/\sqrt{1 + (p/mc)^2} = at/\sqrt{1 + (at)^2}$$
$$a = (qeE/m) \tag{2.9}$$
$$z = [\sqrt{1 + (at)^2} - 1]/a$$

Another simple case is that of motion in a uniform magnetic field. The MKS/SI units for the magnetic field are Tesla (T) or Weber/m², where a Weber is a measure of the flux of the magnetic field B. A Tesla is $1\,N/(Am)$. The permeability of the vacuum is

$\mu_o = 4\pi \times 10^{-7}(\mathrm{Tm})/\mathrm{A}$. The Lorentz force is perpendicular to the velocity, so the momentum magnitude is constant. The direction cosines are α and they will change as the particle bends in the field. The total arc length is the parameter $ds = vdt$, while the radius of curvature is ρ.

In the text, the possibility of a multiply charged particle, such as a helium nucleus, is addressed by assigning a charge qe to the beam particles. However, in detailed calculations, typically only singly charged elementary particles will be considered. The time rate change of the momentum vector is given by the magnetic part of the Lorentz force equation, Eq. (2.10). Since the magnitude of the momentum is constant, the motion can be described in terms of just the direction cosines. The magnetic field is oriented along α_B.

$$d\vec{p}/dt = qe(\vec{v}x\vec{B}), \quad \vec{p} = \gamma m v \vec{\alpha}, \quad \vec{\alpha} = d\vec{x}/ds$$
$$d\vec{\alpha}/ds = qe(\vec{\alpha}x\vec{\alpha}_B)/\rho, \quad \rho = p/(qeB)$$

(2.10)

Choosing the magnetic field to be constant and in the y direction, solutions are easily found by substitution, Eq. (2.11). The momentum vector transverse to the field rotates by an angle ϕ, while the motion in the y direction is a straight line in terms of the arc length s. Initial values of the parameters are labeled by the subscript o. For dipoles in this text, x and z are consistently transverse to the field which is along the y direction.

$$\begin{pmatrix} p_x \\ p_z \end{pmatrix} = \begin{pmatrix} \cos\phi & \sin\phi \\ -\sin\phi & \cos\phi \end{pmatrix} \begin{pmatrix} p_{ox} \\ p_{oz} \end{pmatrix}, \quad \phi = s/\rho$$

$$\begin{pmatrix} x \\ z \end{pmatrix} = \begin{pmatrix} x_o \\ z_o \end{pmatrix} + (\rho/p) \begin{pmatrix} p_z - p_{oz} \\ p_x - p_{ox} \end{pmatrix}$$

(2.11)

$$y = y_o + s(\alpha_{oy})$$

In MKS units, $e = 0.3$ with p in GeV, B in T and ρ in m, which makes the radius of curvature $\rho(\mathrm{m}) = 3.33\,p(\mathrm{GeV})/B(T)$. For example, if $p = 100\,\mathrm{GeV}$ and $B = 1\,\mathrm{T}$, the radius is 333 m. A useful approximation for small bend angles, ϕ, and magnet length, L, is that after $ds \sim dz = L$, $\phi \sim L/\rho$, $y \sim y_o + \alpha_{yo}L$, $x \sim x_o + \alpha_{xo}L + \phi(L/2)$. In terms of a transverse momentum impulse Δp_T, imparted to the particle $\phi \sim \Delta p_T/p$, $\Delta p_T \sim eLB = 0.3\mathrm{B(T)}L(\mathrm{m})$ independent of the particle. For a 1T field with an $L = 1\,\mathrm{m}$ extent, the transverse impulse is $0.3\,\mathrm{GeV}$. In many cases, these approximate results are sufficiently accurate, or give some rapid intuition without calculating the more accurate results.

The solutions for a constant magnetic field are found using the script "Uniform_B.App". The equations of motion are solved symbolically using the Matlab utility "dsolve" for the case when the field is along y and the charge is initially moving along the z axis. A "movie" is made and displayed using the "plot3" utility. The user can vary the "Slider" and change the magnetic field. A screenshot of the output at the end of the "movie" is shown in Figure 2.6.

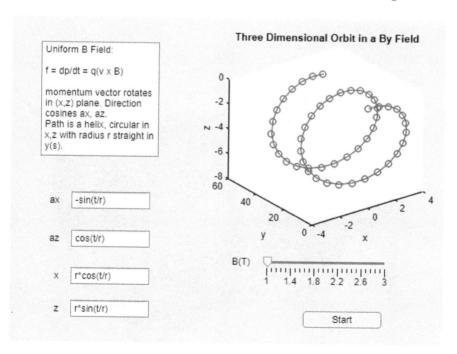

Figure 2.6: Output of the App, "Uniform_B.App". The helical path of the particle is evident. The symbolic solutions are displayed using the "Edit Fields" for symbolic text which are provided.

As was the case for the uniform electric field, some explanation is given using the "Text" box, user input comes from "Sliders" and output is in the form of plots and "Edit Fields", either numeric or symbolic. That format will be uniformly adhered to in this text for most of the Apps. Simpler calculations, such as that shown already for nuclear mean free paths, are performed in Matlab scripts.

As an example of a uniform magnetic field, many experiments use solenoids, while many accelerators use dipole magnets. A solenoid carrying a current I with N total turns of length L and radius a has a magnetic field on the center line of $B = \mu_o(N/L)I[(L/2)/\sqrt{a^2 + (L/2)^2}$. A long solenoid has approximately a purely axial field of magnitude $B \sim \mu_o I n$, $n = N/L$. The dimension of μ_o is (Tm/A). For example, with n in turns/cm and I in A, for $I = 20{,}000\,A$ and 1.6 turns/cm, the field is 4T. The Compact Muon Solenoid (CMS) experiment at the CERN Large Hadron Collider (LHC) is a large 4T solenoid as shown in Figure 2.7.

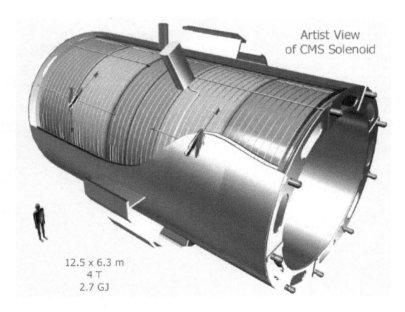

Artist View
of CMS Solenoid

12.5 x 6.3 m
4 T
2.7 GJ

Figure 2.7:   Schematic view of the CMS solenoid. The central field is 4 T and the stored energy is 2.7 GJ.

An example of NR motion in a constant magnetic field is the cyclotron. The field is taken to be a dipole extending over a fixed radius and pointing in the y direction. For NR motion, a charged particle has a rotation frequency in a magnetic field which is independent of particle velocity. The radio frequency, r.f., accelerating voltage can then be at a constant frequency. In general, the rotation frequency is $\omega_o = v/\rho = qeBv/p$ which, for NR motion, is just qeB/m.

Motion in a cyclotron is displayed using the App "Cyclotron" with output given in Figure 2.8. For protons in a cyclotron, NR motion is a good approximation. For a 10 MeV kinetic energy, $T = m_p v^2/2$, cyclotron, $pc = 141$ MeV and in a 1-T field, the radius is 0.47 m. The rotation frequency is 9.6×10^7 rad/s or an r.f., frequency, f, of 15.3 MHz. Linear frequency is f and angular frequency is ω in this text. Using the App, the protons are tracked for a number of turns chosen by "Slider" as is the kinetic energy "kick" supplied by

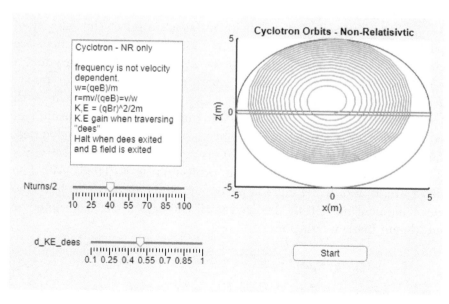

Figure 2.8:   Last frame of the "movie" for NR motion of a proton in a cyclotron. Motion is tracked for a user chosen number of turns and a kinetic energy gain at each traversal of the "dees".

the r.f. when the boundary between the "dees" is traversed. The r.f. is applied only at the boundary between the two "dee"-shaped halves of the vacuum tank and is synchronized to change direction every half rotation. As is usual, the user is encouraged to change the parameters and watch the subsequent "movie". With enough turns or enough acceleration, the particle exits the vacuum and experiments can be performed with the "extracted" beam.

Of course, any actual dipole is of finite length. The field then falls off smoothly with the exact shape set by the specific magnet design. Suppose the main field is B_o directed along the y axis as usual for dipoles. If it falls off to zero over a distance d, then Maxwell's equations require an accompanying field in that region along the z direction, $B_z \sim \pm B_o(y/d)$. In addition, there must also be a field along the x direction which has a factor xy times the second z derivative of the main B_y field. These complications, called "fringe fields", must be addressed in a real accelerator or beam design. Nevertheless, in this text such fringe field effects will be ignored despite their great practical importance.

Uniform crossed E and B fields are used to select the velocity of particles, all of which have the same momentum. Assume there are combined electric and magnetic fields, B_o along x, E_o along y and a particle incident at (0,0,0) along z with momentum p_o. The "force" F is the Lorentz force, Eq. (2.12), which vanishes for particles with the velocity β_o. For reference, a beam with a central momentum can contain charged particles of different masses and therefore velocities, for example, muons of mass 0.113 GeV, pions with mass 0.140 GeV. The kaon mass is 0.494 GeV and proton mass is 0.938 GeV. The determination of which mass a particle is when all particles have the same momentum will be deferred until the next section on detector and beam instrumentation.

$$\vec{F}_{ExB} = qe(\vec{E}_o - \vec{B}_o x \vec{\beta} c)$$

$$\text{if } \beta_o = E_o/(cB_o), \quad \vec{F}_{ExB} = o \tag{2.12}$$

Using $F = dp/dt$ and $p = \beta \gamma m c$, the Lorentz force equation can be worked out. The right-hand side of Eq. (2.12) is $qe[E_o(0,1,0) + B_o(0, \beta z, -\beta y)]$, The left-hand side is: $mc^2\{\gamma(d\beta_x/dct, \ d\beta_y/dct,$

$d\beta_z/dct) + (d\gamma/dct)(\beta_x, \beta_y, \beta_z)\}$. After some tedious calculations, the resulting equations of motion are shown in Eq. (2.13). There is no "force" in the x direction. The electric "force" is along y. The magnetic force, proportional to $1/\beta_o$, is in the y and z direction. The parameter d has dimensions of inverse length but is not a constant due to the presence of the factor γ.

$$d = (qeE_o)/(\gamma mc^2), \quad \beta_o = E_o/(cB_o)$$
$$d\beta_x/d(ct) = d[-\beta_x\beta_y]$$
$$d\beta_y/d(ct) = d[1 + \beta_z/\beta_o - \beta_y\beta_y]$$
$$d\beta_z/d(ct) = d[-\beta_y/\beta_o - \beta_z\beta_y]$$

(2.13)

These equations can only be solved numerically. For both momentum and position, using the Matlab utility "ode45", the solutions are accomplished in the script "ExB_App". There are two choices of B field chosen using a "DropDown" menu. The higher value has a low "drift" velocity when the incident particle is undeflected, and hence the vertical deflection is large. The lower field is more appropriate to a separated kaon beam. At that field, the undeflected value for β can be read off and the incident beam z velocity chosen and matched to the electric field to have an undeflected beam which is plotted. Results of the App are shown in Figure 2.9.

The force due to the electric field is weaker than the magnetic field in the ratio of E/cB for relativistic beams. For example, a 1T B field is three times stronger than a 10^8 V/m E field. For that reason, magnets are typically used to bend and focus beams of relativistic particles, while electric fields are used largely for particle acceleration. The electrostatically mass-separated beams are confined to low-energy applications. The problem can be easily solved approximately assuming small deflections of the incoming beam. Normally, this is a good approximation and is shown in Eq. (2.14) for a region of length L with crossed E and B fields. The fields impart a vertical momentum impulse to any particle that does not have the correct "drift" velocity, β_o.

$$F_y \sim dp_y/dt, \quad dt \sim L/(\beta c)$$
$$d\theta_y \sim (\Delta p_y)/p_o \sim (qE_oL/cp_o)(1/\beta - 1/\beta_o)$$

(2.14)

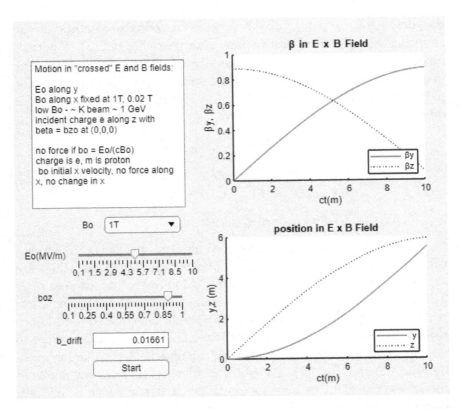

Figure 2.9: Plots of position and velocity for a particle in crossed E and B fields. A "DropDown" menu allows the user to choose a field of $1\,\mathrm{T}$ or $0.02\,\mathrm{T}$. At large magnetic field, the beam will be deflected. At a low field, a specific velocity beam particle will be undeflected. The user should try both options and tune the undeflected velocity.

At the KEK, a particle physics laboratory in Japan, electrostatically separated kaon beam, sextupoles are needed to correct for the chromaticity due to the finite momentum spread in the acceptance of the beam. For the KEK beam: Eo $\sim 5.5\,\mathrm{MV/m}$, Bo $\sim 0.02\,\mathrm{T}$ so that $\beta_o \sim 0.916$. For a beam with central momentum of $p_o = 1.1\,\mathrm{GeV/c}$, pions have a β of 0.992, while kaons have a value of 0.912, or the drift velocity value. Over a length L of $0.4\,\mathrm{m}$, the pions get a vertical angle kick of $0.18\,\mathrm{mrad}$. At a momentum collimator $15\,\mathrm{m}$ downstream of the fields, the pions are deflected by $2.7\,\mathrm{mm}$. Data

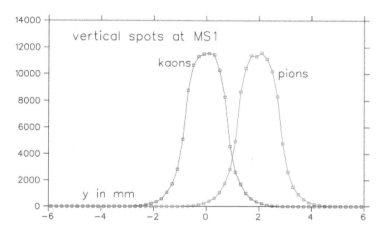

Figure 2.10: Data from the KEK electrostatically separated kaon beam. The actual intensity of pions and kaons are not shown but rather the vertical shapes and separation.

from the beam are shown in Figure 2.10, illustrating the deflection for pions.

2.5. Magnetic Lens — Quadrupole

There are several types of magnets with multipole features beyond a simple dipole. For simplicity, sextupole and higher beam elements are not discussed in this text, although they have great practical importance as correction elements for off-momentum effects. Only the basic quadrupole magnet is explored. The magnetic potential is Φ. The field is derived from the potential. A linear field gradient, dB/dr, is the design goal.

$$\Phi = (dB/dr)xy, \quad \vec{B} = -\vec{\nabla}\Phi \tag{2.15}$$

Maxwell's equations for the quadrupole yield transverse fields $B_y = (dB/dr)x$ and $B_x = (dB/dr)y$. The Lorentz force equations assuming motion with path length $ds \sim dz \sim vdt$ are then: $d^2x/d^2s = kx$, $d^2y/d^2s = -ky$, $k = qe(dB/dr)/p$, and the magnetic fields do no work so that magnitude of the momentum is constant. There is harmonic motion in the focused coordinate, hyperbolic motion in the defocused coordinate. The focal length of a quadrupole of length L

having a field gradient dB/dr is f. A deflection angle for dx/ds is defined to be φ. The units of the parameter k are inverse length squared, while f, the focal length, has units of length.

$$f = 1/(kL), \quad k = qe(dB/dr)/p$$
$$\varphi = L\sqrt{k}$$

(2.16)

A plot of the equipotential lines and the magnetic fields for an ideal quadrupole is derived using the Matlab utilities "contour", "gradient" and "quiver" using the script "Quad_pot_field" and is shown in Figure 2.11. There is no force in the quadrupole center. The forces are toward the origin in one plane, focusing, Q_F, and away from the center, defocusing in the other transverse plane, Q_D. This behavior is the result of Maxwell's equations and is unavoidable. Therefore, a single quadrupole lens cannot focus in both transverse dimensions unlike the situation with optical lenses.

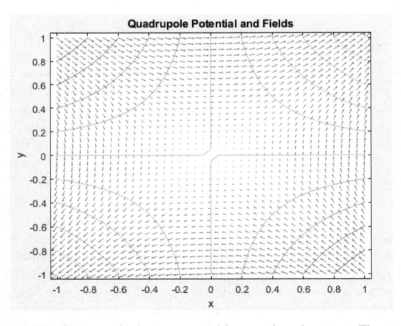

Figure 2.11:   Contours of constant potential for a quadrupole magnet. The arrows are the quadrupole magnetic field indicating both magnitude and direction.

As a numerical example, consider the Fermilab Main Ring (MR) quadrupoles. The length is $L = 2.1\,m$. The field gradient is 24.5 T/m. At 400 GeV, the focal length is $\sim$25 m. In what follows, the MR will be adopted for most discussions of an "accelerator". In this way, the numerical examples are made coherent, which should tie together several distinct topics using the same accelerator parameters.

Maxwell's equations require that if a quadrupole focuses in one transverse dimension, it must defocus in the other one. The equations of motion for small transverse deflections can be solved symbolically using the Matlab utility "dsolve". The output of "Quad_pot_field" is shown in the form of 2×2 matrices in Figure 2.12. The matrices are defined to multiply vectors of transverse coordinates and "velocities", $[x, dx/ds]$ and $[y, dy/ds]$ for a focusing and defocusing quadrupole, respectively. This matrix form is well matched to Matlab as a vector language. Extensive use of the matrix capabilities of Matlab will be made in the later discussions and expositions. For small values of φ in

```
Bx =

-dBdr*y


By =

-dBdr*x


MF =

[                  cos(ph),  sin(ph)/k^(1/2)]|
[ -k^(1/2)*sin(ph),                 cos(ph)]


MD =

[                 cosh(ph),  sinh(ph)/k^(1/2)]
[ k^(1/2)*sinh(ph),                 cosh(ph)]
```

Figure 2.12: Fields for a quadrupole and the matrices representing passage through a quadrupole with a gradient and length L defining the parameters $k = qe(dB/dr)/p$ and $\varphi = L\sqrt{k}$.

Eq. (2.16), the quadrupole can be thought of as causing a transverse momentum "kick", $\Delta p_{TQ} = eL(dB/dr)x$. For the MR quadrupole example, the "kick" is $0.154 \, \text{GeV/cm}$ of off-axis displacement, which changes the angle, at $400 \, \text{GeV}$, by $0.385 \, \text{mrad}$.

It requires two quadrupoles to supply overall (x, y) transverse focusing. The simplest system is a doublet of Q_F and Q_D acting together. Such a configuration often appears in external beams and is the basic unit cell of many accelerators. Those facts justify a more detailed exploration of the quadrupole doublet now before taking up specific details respecting beam lines and accelerators since they are a common element of both. The difference is that in a beam line, the doublet often satisfies a focal constraint, while for an accelerator, the doublet gently makes the beam oscillate with small displacements, called betatron oscillations. In addition, in a beam line each particle traverses a magnet only once, while in an accelerator, multiple traversals are, in fact, the goal.

2.6. Quadrupole Doublet — (x, y) Focus

Very often, a thin lens approximation can be used, especially for a first approximate solution. In that approximation, the quadrupole simply imparts a change of transverse direction to a particle with $M(2, 1) = + - (1/f)$, $M(1, 1) = M(2, 2) = 1, M(1, 2) = 0$. This approximation is used extensively in what follows, although a more correct doublet problem is treated once. A free space or "drift" of length L has $M(1, 1) = M(2, 2) = 1$, $M(1, 2) = L$, $M(2, 1) = 0$. A quadrupole of length L often also uses $M(1, 2) = L$.

A doublet consisting of a Q_F quadrupole followed by a field — free, or drift space, followed by a Q_D quadrupole is used. A quadrupole is assigned F or D in regards to the x direction. Initial conditions are specified at a point upstream of the Q_F quadrupole, either a point, $[0; 1]$, $(x_o = 0)$, or a parallel beam, $[1; 0]$ $(dx_o/ds = 0)$. Symbolic solutions for constraints are obtained by multiplying the matrices for the drift spaces and the quadrupoles together and applying constraints on the "focal condition" location downstream of the doublet. The focal conditions applied to the overall transfer matrix, M_T, can be a parallel to point focus, $M_T(1,1) = 0$, or a point to point beam, $M_T(1,2) = 0$ or a point to parallel, $M_T(2,2) = 0$. In the symbolic solutions, do is the distance to Q_F, d is the distance between Q_F and Q_D, and dc is the distance from Q_D to the location of the constraints. The Matlab utility "solve" for algebraic solutions is used to solve constraints in x and y for the two quadrupole focal lengths.

The App "Doublet_Thin_Symbolic" finds the solutions symbolically with results shown in Figure 2.13. The type of constraint is chosen using the "DropDown" tool. The symbolic solutions for the

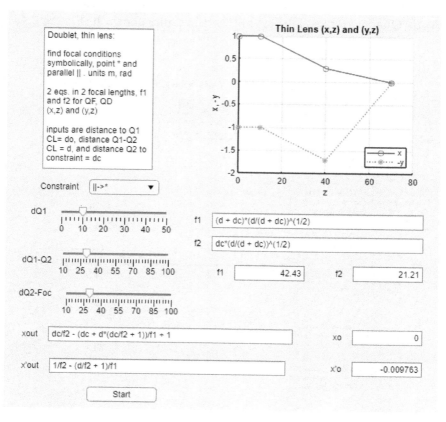

Figure 2.13: Output of the script "Doublet_Thin_Symbolic" for a particular set of Slider and Constraint choices.

two focal lengths are displayed and then the numerical values are also shown using the "EditField" tools. Using the "Sliders" to set the values for the three distances, a plot of the solutions in both x and y is made. The beam conditions for x and dx/ds at the constraint location are also shown both symbolically and numerically, again with the "EditField" tools. Since the doublet is the simplest layout that can provide a focus in both x and y, it is well worth studying. The user is accordingly encouraged to play with the constraints and the numerical values of the three distances.

Normally, a thin lens approximation can be made. In any case, it is a useful first approximation to the final, thick lens, calculation. The approximation, compared to an exact calculation is explored in the App "Quad_Doublet_Thick_Thin" with results shown in Figure 2.14.

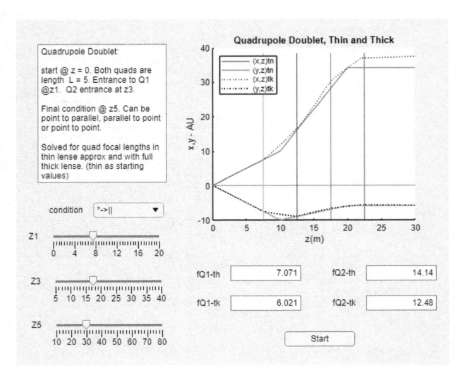

Figure 2.14:   Output of the script "Quad_Doublet_Thick_Thin" comparing a thin lens solution to the full thick lens solution.

The full matrices shown in Figure 2.12 are used. The Matlab utility "fminsearch" is used to numerically solve the nonlinear problem of adjusting quadrupole strengths to attain the focal conditions. For example, $|M_{Tx}(2,2)+M_{Ty}(2,2)|$ is minimized for the point to parallel condition, as before where M_T is the full transfer matrix through the five beam elements. The "fminsearch" utility minimizes an arbitrary user-defined function of several variables. The constraints are the same as previously imposed in the thin lens approximation, with numerical inputs for the distances as before. The numerical outputs are the focal lengths for both the thick and thin lens solutions and the plots of the beam in the two cases. It should be clear that the thin lens is a good starting value for the focal lengths, which is important since nonlinear fitting can be unstable if started far from the actual solution.

This problem is more complex. It is solved using several functions with communication using the "function" results and also using the "global" Matlab feature to define common variables. The subsidiary "functions" are "Doublet_Thin" and "Doublet_Plot" to find and plot the thin lens solutions. The functions "Quad_Matrix" and "app.Doublet_Fit" are used to define the full quadrupole matrices and to define the function for "fminsearch" to minimize. The focal lengths are systematically smaller in the full thick lens treatment than in the thin lens approximation. The user can appreciate the added complexity that arises in the thick lens case by the difference in the length of the code in the two Apps.

The motion of charged particles in a uniform electric field, a uniform magnetic field (momentum selection) and a combined, crossed, uniform electric and magnetic field (velocity selection) has been introduced. Also introduced were the magnetic quadrupole lens and the use of two such lenses to obtain overall focusing in both transverse coordinates simultaneously. In future, it will be assumed that a beam of charged particles has been produced by bombarding a target, for example, and then capturing the beam using quadrupole lenses. A selection of the momentum of the beam can be made using dipole, or bending, magnets, and collimating the resulting chromatic spectrum of angles induced by the dipoles. The beam has a finite acceptance in momentum, $\delta = dp/p$, and in transverse coordinates and velocities, x, dx/ds, y, dy/ds. The beam may also contain a mix of particles of different masses. The next section begins to explore the tools which are deployed to understand and control such beams.

2.7. Special Relativity

In classical physics, energy and momentum are separately conserved. In special relativity, SR, energy and momentum are components of a four-dimensional vector. We have taken kinetic energy to be $T = p^2/2m$, $p = mv$. In SR, the energy for a material particle is ε with $\varepsilon^2 = (mc^2)^2 + (cp)^2$, so that at low momenta, $\varepsilon \sim mc^2 + p^2/2m$, the rest energy plus the NR kinetic energy. In general, $\varepsilon = \gamma mc^2$ with $\gamma = 1/\sqrt{1 - \beta^2}$. One can approximately think of the effective mass gaining a factor of γ due to motion. For momentum $p = \gamma m \beta c$ and again $m- > \gamma m$ is a useful way to think of momentum in SR. These expressions have already been used earlier in the text. In this section, some general SR topics are gathered together.

In the case of photons SR assigns a momentum and energy proportional to the wave vector k and the circular frequency ω, respectively. For a photon in a reference frame carrying energy and momentum ε_* and p_* with angle θ_*, the corresponding quantities in a frame moving with respect to the $*$ frame with velocity βc along the z axis are shown in Eq. (2.17). These transformation factors will be encountered often in the following text:

$$\varepsilon = \hbar \omega, \quad \vec{p} = \hbar \vec{k}$$

$$p_\perp = p_{\perp *}$$

$$p_{||} = \gamma \varepsilon_* (\beta + \cos \theta_*) \tag{2.17}$$

$$\varepsilon = \gamma \varepsilon_* (1 + \beta \cos \theta_*)$$

Consider the fully relativistic two body elastic scattering kinematics, as this will be needed later in the text. By elastic it is assumed here that new, additional, particles are not produced. The App used is "SR_kin_2_Body_El" and it is limited to two-body scattering with no production of new particles. The App is written in the laboratory frame where particle #2 is initially at rest and particle #1 is incident along the z axis. The two-particle energy and velocity in the center of momentum frame are calculated and displayed using the "EditField" tool. That particular frame is the frame where the total momentum of the two particles is zero. Two sets of kinematic plots are available, chosen by using the "Switch" capability with the first set shown in Figure 2.15. The mass m_1 is defined to be 1; target mass m_2 and incident momentum p_1 are chosen by "Slider" and are in energy units for simplicity. By energy units, one means pc and mc^2 are used and a shorthand is adopted with mass and momenta expressed in GeV.

By playing with the App, the user can find that for target masses equal to the projectile mass, the outgoing angles go from 0 to a

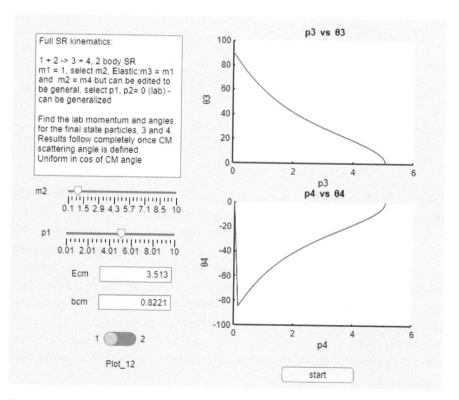

Figure 2.15: Output of (p, θ) correlations for the SR kinematics for $1+2 \rightarrow 1+2$ (elastic) processes.

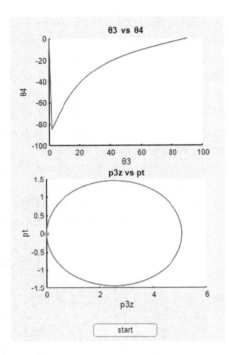

Figure 2.16:   Second set of kinematic plots in the special case of equal mass target and projectile. For small angle scattering, the projectile exits with a small angle and the target recoils at 90°.

maximum of 90 degrees. For target masses greater than the projectile mass, the outgoing particles can go backwards, bouncing off the heavy target. For target masses less than the projectile mass, the outgoing projectile always goes forward as does the target. The target exits at smaller angles as the projectile mass increases with respect to the target. A second set of plots is shown in Figure 2.16. The angular correlation of the outgoing particles is displayed along with the locus of points for the outgoing projectile in the special case that $m_2 = m_1$. The user is encouraged to use the "Sliders" and see how the scattering process depends on the target mass and the projectile momentum.

Chapter 3

Detector and Beam Instrumentation

"By allowing the positive ions to pass through an electric field and thus giving them a certain velocity, it is possible to distinguish them from the neutral, stationary atoms."

— **Johannes Stark**

"The application of a strong magnetic field enables the measurement of the energy of the most penetrating particles to be carried out, and the method may be capable of still further extension and improvement."

— **Victor Francis Hess**

3.1. Photoelectric Effect

Detectors are needed to measure the positions, angles, momenta, and perhaps mass of the particles in an external beam line or in an experiment. To that end a variety of physical processes are deployed. They almost all rely on the electromagnetic interactions of the particles because electrons are freed from atoms fairly easily. First, the photoelectric effect is explored. It is the dominant process at low photon energy, possessing the largest cross-section. Its use is ubiquitous in everyday devices in addition to the specific instrumentation herein considered.

Consider the interaction of a low-energy photon with an electron bound in an atom with atomic number Z and atomic weight A. The photon is characterized by frequency f, wavelength λ, wave number k, velocity c, and circular frequency ω. Numerically, $\lambda = 1240 \text{ eVnm}/\varepsilon_\gamma(\text{eV})$. An atom with Z electrons in different quantum states labeled by quantum number n have different energies, ε_n which are approximated using the hydrogen atom solution scaled up by Z^2.

Bound states have negative energies. Such an electron can be freed, with a positive final energy, by an energetic photon. The process is totally inelastic in that the photon is absorbed by the atom, not scattered.

$$\lambda f = c, k = \omega/c, \omega = 2\pi f, \lambdabar = \lambda/2\pi = c/\omega = 1/k$$
$$\hbar kc = \hbar\omega > -\varepsilon_n = m_e c^2 (Z\alpha)^2/(2n^2) \tag{3.1}$$

In copper, the photoelectric effect dominates over the effect of Thomson scattering for photon energies below about 0.1 MeV. The cross-section falls with photon energy approximately as the 7/2 power. At low energies the cross-section is of order megabarns, comparable to the geometric cross-section of the atom. For the state $n = 1$ in copper an approximate photon energy threshold of 11.4 keV is needed to free the electron. A plot of the cross-section of a photon on copper as a function of photon energy appears in Figure 3.1. At photon energies above about 0.1 MeV Compton

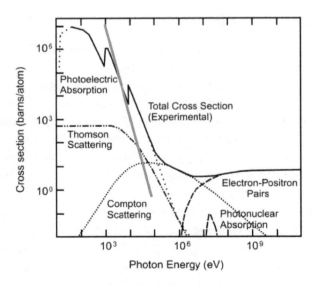

Figure 3.1: Cross-section for photons scattering on copper as a function of photon energy. The red line is an approximate photoelectric behavior, with the cross-section falling as the inverse 7/2 power of the energy.

scattering dominates the cross-section, with electron–positron pair production the dominant process above about 10 MeV.

An approximate expression for the photoelectric cross-section appears in Eq. (3.2). The scaling of the cross-sections goes approximately as Z^5, α^6 and $\omega^{-7/2}$. The structure in the cross-section seen in Figure 3.1 has to do with specific quantum states in copper, where the $n = 1$ and 2 states are evident at ~11.4 and 2.8 keV. The result for the approximate cross-section, Eq. (3.2), for a 10 keV photon on copper is 0.21 Mb which is in rough agreement with the data, although detailed calculations are needed to make quantitative predictions.

$$\sigma_{pe} \sim (8\pi/3)(4\sqrt{2})Z^5(\alpha^3 \lambdabar_e)^2(m_e c^2/\hbar\omega)^{7/2} \qquad (3.2)$$

Note that the cross-section has the correct dimensions, the reduced electron Compton wavelength squared, times dimensionless coupling, Z, α, and numerical factors. The deBroglie wavelength is sometimes encountered in other texts. It is the wavelength associated with the matter wave of a particle with momentum p and is defined to be $\lambda_{dB} = h/p$. The final electron NR kinetic energy, $T = mv^2/2$, is approximately the incoming photon energy less the binding energy, ε_n. The threshold photon energy is $-\varepsilon_n$ which depends on the quantum state of the atomic electron.

The angular and kinetic energy, T, distribution of the free electron with respect to the incident photon is shown in Eq. (3.3). The photoelectron has a dipole $\sin^2 \theta$ factor, but is emitted preferentially near the direction of the incident photon due to the effect of the denominator. This forward behavior increases as the kinetic energy of the electron increases. This type of behavior occurs in many processes some of which will be covered later. Note that this definition of kinetic energy is only approximately equal to $mv^2/2$ for NR motion and is not applicable for the energies shown in the figure. The differential cross-section is plotted using the script "Pe_effect_dsig_dO" and the "surf" utility to display the two-dimensional surface as shown in Figure 3.2. The forward peaking as T increases is a prominent feature.

$$d\sigma_{pe}/d\Omega \sim [\sin^2 \theta/(1 - \beta \cos \theta)^4], \quad T = \varepsilon - mc^2 \qquad (3.3)$$

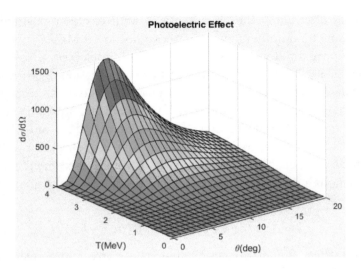

Figure 3.2:   Plot of the angular distribution of electrons as a function of the kinetic energy of the electron.

3.2. Photomultiplier Tube (PMT) and Scintillator

The photoelectric effect is employed in a widely used detector element, the photomultiplier tube (PMT). A scintillator is a material which is ionized by the passage of a charged particle, as described later in Section 3.8. That ionization energy excites "fluors" which rapidly de-excite and emit light in the visible region by first emitting in the ultraviolet which is then absorbed by a secondary fluor which subsequently emits visible light. At these visible wavelengths, the scintillator light is little absorbed by the scintillator itself and can be detected using a PMT. A detector needs to be transparent to its own emissions. A schematic plot of the absorption and emission spectrum of a typical fluor is shown in Figure 3.3. Some popular fluors are PTP whose emission spectrum peaks at 440 nm and POPOP at 420 nm. The decay times are typically quite fast, of order a few nanoseconds. Typical base plastics are polystyrene and poly-vinyl-toluene (PVT), although several other plastics have been used. The processes before emission of visible light are very localized as indicated in Figure 3.3 which means the visible light can be a measure of the position of the incident charged particle.

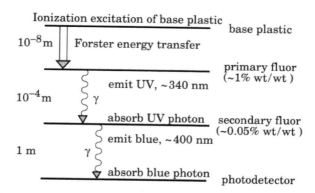

Figure 3.3: Schematic diagram of the steps occurring in a typical scintillator from the initial deposit of ionization energy to the final emission of blue light for which the scintillator is reasonably transparent, allowing the light to exit to a transducer such as a photomultiplier.

The wavelength shifting (WLS), between absorption and emission is necessary to place the photon wavelength in a region where the plastic is reasonably transparent. The shift can be by ~50 nm or ~0.28 eV. A typical plastic scintillator may create 50,000 visible photons per meter of particle traversal. For a 1-cm thickness scintillator with a light collection efficiency of 10% incident on the PMT, there are 50 photons to detect. A plot of typical absorption and re-emission spectra of a typical fluor is shown in Figure 3.4.

The visible photons then exit the scintillator and strike a photocathode. This is a thin layer of special material with a large photoelectric cross-section and a low value of the binding energy so as to efficiently convert visible photons. A thin transmission cathode and a vacuum are required because the photoelectrons are not very energetic and can be easily absorbed or deflected before collection as a current. A transmission photocathode is favored by the forward angular distribution of the photoelectrons. The electron quantum efficiency (QE), or the number of photoelectrons observed per incident photon is typically at most about 20%. This is a basic limitation in the use of the PMT.

The resulting emitted electrons are electrostatically collected in the vacuum of the PMT and the signal is amplified in "dynodes"

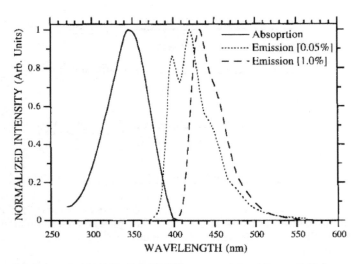

Emission spectra of a highly diluted (0.05%), pure sample in solution, and 1.0% fluor sample in PVT matrix, i.e., plastic scintillator form.

Figure 3.4:   Spectra of the absorption and re-emission of photons for a typical primary and secondary fluor. The details of the emission spectrum depend on the concentration of the fluor in the base plastic as indicated in the figure. The two spectra should not overlap significantly in order to avoid re-absorption.

which impart gain to the electron current using secondary emission and multiplication. A schematic of the dynodes is shown in Figure 3.5. The dynodes each have an overall gain g. An "n-stage" tube then has total gain g^n. For a 14-stage tube with dynode gain of 3 the overall gain is 4.8×10^6. With a 20% QE there are 10 photoelectrons. For a 14-stage tube if the electrons are delivered in ~5 ns, there is a signal of 1.54 mA into 50 ohms or 76 mV. Such a signal is normally easily distinguished from noise sources. A more robust PMT signal is displayed in Figure 3.6. Clearly these devices can measure small signals of a few photons with a temporal accuracy of order nanoseconds or less. However, the PMT is a relatively old workhorse technology and new devices are now in use. In particular the spatial size of the scintillator and the PMT itself preclude spatial resolutions less than a centimeter. In addition they are fragile and require large voltages to be supplied to the dynode chain.

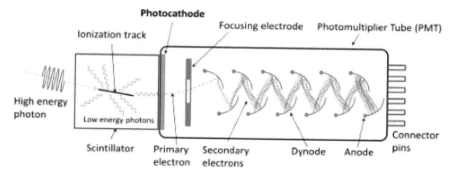

Figure 3.5: Schematic diagram of a PMT showing ionization in the scintillator converted into optical photons. The photons strike the photocathode and the emitted photoelectrons are amplified and collected on the anode output as a current pulse. As with all electron signals, the anode voltage is positive with respect to the cathode.

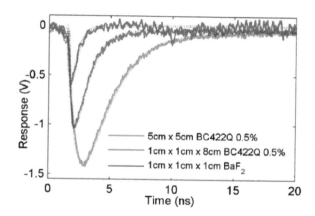

Figure 3.6: Current pulse from a PMT illuminated by different scintillators. The rise time is a few nanoseconds as is the pulse width.

3.3. Magnetic Shielding

The PMT is very sensitive to stray magnetic fields because the emitted photoelectrons have very small kinetic energies, of order eV, and would curl up in the field. Magnetic "shielding" is often required to preserve the electron trajectories in a PMT dynode structure and maintain the needed gain. A typical method for protection is to surround the PMT with a material of very high magnetic permeability, μ compared to the vacuum value of μ_o. This material concentrates any external magnetic field and thus reduces the field inside the shield. Electric fields are more easily shielded because metals surrounding an instrument exclude the fields — a "Faraday cage" — especially at high frequencies as will be explored later in a discussion of the "skin depth" in Section 3.18.

For an external magnetic field B_o, the region inside a spherical shell of inner radius a and outer radius b is reduced by a large factor if μ is large. An approximate relationship is shown in Eq. (3.4) which shows that the reduction in field scales as $1/\mu$. The exact solution follows from applying the magneto-static boundary conditions. These are illustrated using the App "magnetic_shield" with a specific output shown in Figure 3.7.

$$B_{in}/B_o \sim 9/[(2\mu/\mu_o)(1 - (a/b)^3)]. \tag{3.4}$$

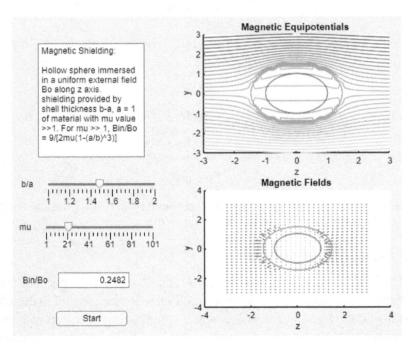

Figure 3.7:   Plot of the magnetic potential and the magnetic field for a spherical shell of inner (red) and outer (green) radius (a and b) and permeability ratio μ/μ_o.

The App uses the solutions for the magnetic potential with b/a and μ set by "Slider". The potential is plotted using the Matlab utility "contour" and the magnetic field is plotted using "gradient" and "quiver". The numerical value for the reduced field inside is displayed using an "EditField". The user can adjust the parameters in order to see the effect on the interior field.

3.4. Silicon PMT

Recently solid state devices called Silicon PMT or "SiPM" have been developed which can replace a PMT. They operate at much lower voltages than the typical kilovolt voltages needed for a PMT and are not vacuum devices, thus avoiding the fragility of a PMT. In addition, their QE is much larger than that of a PMT, perhaps three times higher. They are small and immune to the issues with external fields that often require magnetic shielding for PMT. Their speed is comparable to a PMT and the detector elements can be made into small pixels thus affording much better position resolution. The cost per independent pixel is also less than the cost of a PMT.

An incident optical photon is photoelectrically absorbed by the SiPM. This causes a local avalanche in a high electric field region, as in a Geiger tube, which gives the SiPM a large gain comparable to that of the PMT using several sequential dynodes. The specific pixel then must recharge, but the device has many pixels, so that this dead area is a small fraction of the SiPM surface area. Thus, the probability of a photon not causing a SiPM signal is small. The QE is very high with respect to a PMT. Since the basic statistical signal fluctuation is due to the limited number of photoelectrons, this is a non-trivial advantage.

A schematic of a SiPM is shown in Figure 3.8. The incident photon is absorbed, liberating an electron which forms an avalanche by impact ionization in the high electric field of the SiPM. That discharges the specific pixel, which is inoperative until recharged. All the struck pixels are read out. Later in the text solid state detectors will be explained in much more detail, Section 3.17, and the avalanche

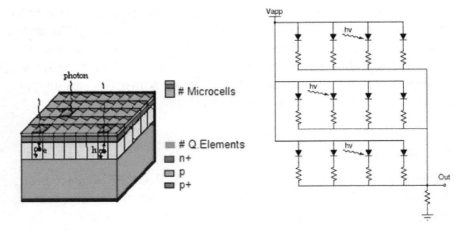

Figure 3.8:   Schematic diagram of the pixel structure of the SiPM, which operates in Geiger mode.

gain mechanism will be explored in the discussion of proportional wire chambers (PWCs) in Section 3.16. For now, the aim is simply to compare to a classical PMT and highlight the advantages of a SiPM.

Such devices are finding wide application in high-energy physics experiments. Indeed, the good spatial and temporal resolution, ruggedness, and low voltage requirements make the SiPM attractive in many areas. For example, PET/CT scanners with SiPM readout instead of PMT are finding their way into the medical mainstream. In general, solid state devices are becoming favored in many fields of technology for similar reasons. The quantum efficiency for a SiPM is several times larger than the PMT. The fluctuations in gain for a SiPM are therefore smaller for the same number of incident photons. This property allows the SiPM to resolve the number of photoelectrons as seen in Figure 3.9. Such resolution is not available in a PMT and allows the SiPM user to define a calibration signal due to a single photoelectron.

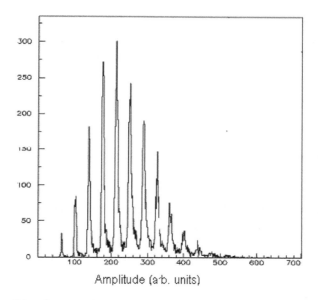

Amplitude (arb. units)

Figure 3.9: Distribution of the number of struck SiPM pixels illuminated by a light source, equally spaced in amplitude. The individual peaks arise for a variable number of emitted photoelectrons. The Poisson distribution mean is about five photoelectrons with standard deviation ~2.2 while a PMT would wash out such detailed structure.

3.5. Time of Flight — β

The PMT or the SiPM have very short rise times, of order ns, or 10^{-9} s. If the signal to noise of the detectors is sufficiently large and the statistics of the pulse is good, the time resolution of the signal can be better than the rise time, perhaps at the 100 ps level. The time of flight (TOF) of a charged particle between two detectors a distance L apart measures the velocity of the particle, $v = \beta c$. If the particles are in a beam which has been momentum selected using dipoles and collimators, then particles of different mass, for example pions, kaons and protons, have different TOF values. The difference in TOF for two different masses decreases with the inverse square of the beam momentum, which makes the TOF technique limited to fairly low beam momenta. For example, with $L = 10$ m, $cp = 3$ GeV the K–π difference in TOF is 0.43 ns, while the difference for p–π is 1.58 ns. These time differences are comparable to the time resolution of a PMT.

$$\text{TOF} = L/(\beta c)$$

$$d(\text{TOF}_{12}) \sim (L/2c)[(m_1^2 - m_2^2)/(p/c)^2]$$

$$(3.5)$$

The TOF measured in a beam of fixed momentum is explored in the App "dEdz_TOF". The output is shown in Figure 3.10 for the TOF of pions, kaons and protons as a function of beam momentum. The distinct curves collapse together as all particles approach a velocity c in SR as the energy increases. Note that any spread of

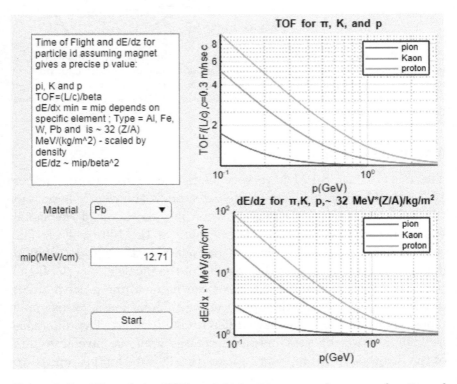

Figure 3.10: Plot of the TOF and ionization energy loss as a function of momentum for pions, kaons, and protons.

momenta in the beam, dp/p, is ignored here. That spread would widen the curves. Clearly, the TOF technique works well only for low momentum beams.

As will be shown later, a singly charged particle deposits an ionization energy in material which scales as the inverse square of the β factor for that particle when it is NR. All such particles deposit the same energy if they are relativistic, the so-called "minimum ionizing particle" or "mip". In Figure 3.10, there is a "DropDown" menu that allows the user to choose several materials. The energy deposited by a "mip" is displayed in the "EditField". The plot of energy loss per unit length of material traversed also shows a convergence of all charged particles as the velocity approaches c. Both methods of measuring velocity have a similar behavior. Using the magnetic field of the beam dipoles to limit the momentum spread using collimation and using either the TOF or ionization to measure velocity, the mass of particles produced in a collision can be identified.

Data from the ALICE experiment at the LHC on the TOF as a function of momentum is shown in Figure 3.11. At low momentum there is a clear separation of the different particle types at a fixed momentum. Errors on the measurements due to both the momentum error and the TOF error give the curves a finite width. All the curves

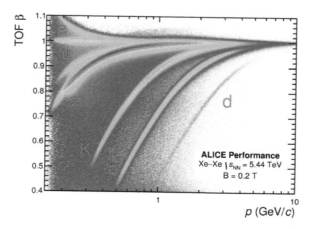

Figure 3.11:   Data on the TOF as a function of momentum in LHC collisions for p, K, π, μ, e and even deuterons.

converge at high momentum values. These two techniques both measure the velocity and thus, at fixed momentum, the mass. For this reason, they are compared here rather than in distinct discussions.

3.6. Cerenkov Counters — β

The use of Cerenkov radiation allows for the extension of particle identification (ID) to higher values of velocity than the use of ionization or TOF. A charged particle emits radiation if it exceeds the velocity of light in some medium. The medium is here defined by an index of refraction n, assumed to be real. The Cherenkov angle, θ_c, is the angle of the photon with respect to the particle direction. Light is not emitted below a threshold velocity, $\beta_{\text{th}} = 1/n$. These properties are made visible using the App "Mach_Cone". The user selects the velocity of the particle in the medium, either below or above threshold. When above threshold the outgoing spherical waves align into a "Mach cone" whose angle increases as the velocity

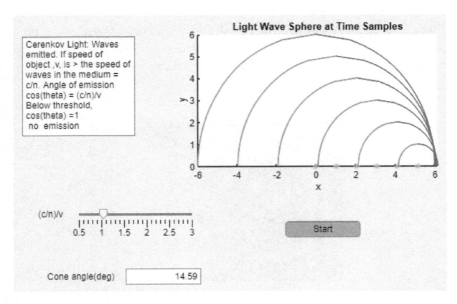

Figure 3.12:   Locus of points from six distinct light emissions (green dots). The forward light (blue) and backward light (red) are both plotted and above threshold the forward light adds up into a cone with a ∼14.6 degree angle.

increases. A specific output appears in Figure 3.12. In this case, $\cos\theta = v_s/v$ where v_s is the velocity of signal propagation $= c/n$. A "movie" is made of the outgoing waves for six distinct points of emission.

To deal with large value of β, gases with indices of refraction near one are used. Simple expressions are given for gases with index ~ 1 and for particles with $\beta \sim 1$ in Eq. (3.6). There is a threshold value for γ, γ_{th}, appropriate for the cases where β is near 1. General relationships for waves in a medium are $v = c/n$, $k = n(\omega/c) = \omega/v$. In vacuum the value of ε_o in MKS units, Farad/m, and of μ_o in units of N/Amp^2 are such that $c = 1/\sqrt{\varepsilon_o\mu_o}$. There are two modes of Cerenkov counter operation. The index can be chosen by selecting the gas pressure. There is a threshold mode — light or no light. There is also a differential mode — measuring the emission angle to select the mass.

$$n = c\sqrt{\varepsilon\mu}, \ \cos(\theta_c) = 1/(n\beta)$$
$$n = 1 + \delta n, \ \beta = 1 - \delta\beta, \ \theta_c \sim \sqrt{2(\delta n - \delta\beta)}$$
$$\theta_c^2 \sim 1/\gamma_{th}^2 - 1/\gamma^2, \ \gamma_{th}^2 = 1/(2\delta n), \ \gamma^2 = 1/(2\delta\beta)$$
$$\theta_c^2|_{max} = 2\delta n$$

$$(3.6)$$

Cherenkov light is emitted uniformly over all frequencies and at all points along the path, defined by z, as long as the index is constant and real which limits the frequency range. The assumption is that the radiating particle loses only a small amount of energy since there are only a few photons each with a few eV energy which are radiated.

$$d^2N_\gamma/d(\hbar\omega)dz = (\alpha/\hbar c)(\mu\varepsilon_o c^2)\sin^2\theta_c$$
$$\sim 3.65 \times 10^4 \sin^2\theta_c/(eVm)$$

$$(3.7)$$

The number of photons scales as the counter length, L, and as the photon energy window for detection, $N_\gamma = (\alpha/c)\sin^2\theta_c\Delta z\Delta\omega$. The spectrum is flat in photon energy as long as the medium is transparent to such photons. Extracting the light requires UV transmission if possible, as long as the emitting medium is transparent. For example, expensive quartz windows are often used if required. A few indices of

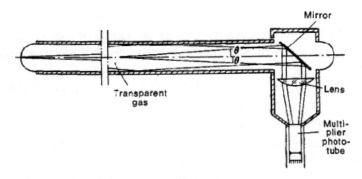

Figure 3.13:   Schematic of a gas filled Cerenkov counter installed in a beam line.

refraction for gases and solids are displayed in Appendix D. A typical Cerenkov counter installed in a beam line is shown schematically in Figure 3.13. The light is sent to a PMT after traversing the gas, being bounced off a mirror, focused by a lens, and exiting through a window, all of which should be transparent over a wide frequency range. The counter is operated in threshold mode with or without a PMT signal.

In order to get an intuition about gas filled Cerenkov counters, an App, "Beam_ID_Crenkov" can be invoked. A specific output appears in Figure 3.14. The user selects the index of a gas. The emission angle for protons, kaons and pions is plotted as a function of momentum of the particles. The maximum angle and the threshold γ factor is displayed via "EditFields" as well as the number of photons per eVm. Clearly, although higher momenta can be distinguished in mass or velocity, the number of photons available limits the detection efficiency and requires some attention. One can note also that a finite beam angular divergence makes the cone angle determination difficult for very high energy operation due to the small Cerenekov angles.

A numerical example for a 50-GeV beam is useful to see the issues. Assume that the beam is negative and that labelling each particle as a pion or kaon is the role of the Cerenkov counter. For pions and kaons the value of $\delta\beta$ is 3.9×10^{-6} and 4.7×10^{-5}, respectively. An index of $\delta n = 10^{-4}$ is chosen so that both pions and

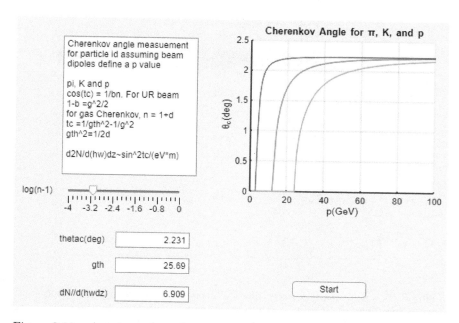

Figure 3.14:   App to explore the use of gas filled Cerenkov counters for indices of refraction as low as 1–0.0001. At a fixed momentum a particle may not emit light or may emit at a fixed angle.

kaons radiate. The Cerenkov angles are 13.9 mrad and 10.1 mrad. For a 10-m long counter with a range of photon energies of 1 eV the number of photons emitted is 70.1 and 37.5. With a 20% quantum efficiency of a PMT, the number of photoelectrons is 14 and 7.5. The statistical fluctuations in the number of photoelectrons is large, ~ 4, which can be ameliorated by reconstructing the angle of emission of the photons as well as the number of photoelectrons. That device is called a Ring Imaging Cerenkov Counter (RICH).

Data shown in Figure 3.15 for a measured Cerenkov angle as a function of beam momentum indicates the difficulties for beam momenta above about 50 GeV for pion/kaon seperation. The maximum angle of about 30 mrad shows that the index of refraction of the gas used is $\delta n \sim 4.5 \times 10^{-4}$. A very parallel beam region is required along with a limited momentum acceptance. Because the Cerenkov counter measures β, these issues are intrinsic. Later in the

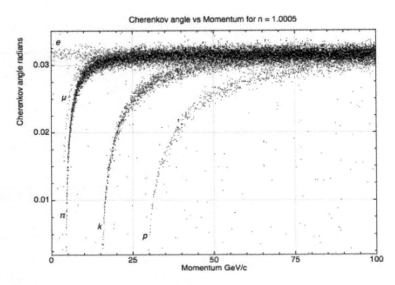

Figure 3.15:   Data using a beam line Cerenkov to distinguish pions, kaons and protons in a positive beam as a function of beam momentum. For pion and kaon separation the device begins to fail at about 50 GeV beam momentum.

text, Sections 3.9–3.11, processes that depend on γ will be discussed. These processes can be used to extend the range of momenta where particle ID is possible.

3.7. Scattering

Scattering of particles is a fundamental process. Indeed, different forces between particles lead to different scattering angular distributions so that scattering experiments give information about the forces acting on the particles. Consider the scattering of a charge qe, $q = 1$ for protons, by a fixed atom of charge Ze when the incident particle has an impact parameter b. The simplifying assumption is made that the scattering center is extremely heavy, so that it remains fixed in space during the scattering process. The impact parameter is the perpendicular distance between the incident particle velocity and the scattering center at large distances between them. Large angle scattering will occur when the target is the point-like nucleus rather than the much more widely dispersed atom.

For example, the Coulomb force for an undeflected incident charge, $F(b)$, the approximate time of interaction, $dt(b)$, and the scattering angle, $\theta_R(b)$ are, assuming a small deflection angle, shown in Eq. (3.8). The potential energy seen by the projectile is defined to be $U(b)$ here, while the NR kinetic energy is T. The kinematics are non-relativistic (NR). Fully relativistic scattering will be discussed later in the text. In small angle approximation the scattering angle is simply the potential energy at a radius of b divided by the incoming kinetic energy.

$$F(b) = (qZe^2/4\pi\varepsilon_o)(1/b^2) = a/b^2, \ \Delta t(b) = 2b/v$$
$$\theta_R(b) = \Delta p_\perp(b)/p = F(b)\Delta t/p = U(b)/T = (a/b)/T, \quad (3.8)$$
$$T = mv^2/2$$

There are five distinct b values shown in Figure 3.16 generated by the App "Impact_DCA_App". The App allows the user to look at scattering for different force laws having different powers of b, both attractive and repulsive, chosen using the "Switch" utility. The power law is chosen by the "DropDown" menu utility. The numerical solution is obtained using the Matlab utility "ode45" and is plotted as a movie in time. The symbolic solution for the distance of closest approach, DCA, is produced when a closed form solution can be found for n of two or three. Since the forces are central, angular momentum conservation is used in the computation of the symbolic result for the DCA. That result appears in the "EditField" if it exists, otherwise "no_sol" is written out. Note that the scattering center, shown in the plot as a red point, does not recoil, and the process is fully elastic. The symbolic solution can be unavailable. Even in that case the numeric "ode45" solutions are still instructive. Playing with the options and viewing the resulting movie can be quite useful.

Although the previous example is instructive it is unrealistic. The incoming beam cannot be aimed at a specific nucleus, so the differential cross-section, $d\sigma$, is simply the transverse area element for a given impact parameter since all transverse areas are equally probable. For azimuthally symmetric forces, $d\sigma$ is proportional to bdb.

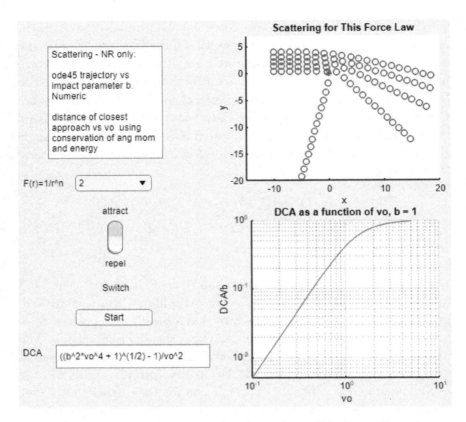

Figure 3.16: Specific output of the App, "Impact_DCA_App". The user can choose the power law for the force and whether it is attractive or repulsive. For large values of v_o the DCA is approximately b and the deflection is small.

Since each b has a unique scattering angle, θ, the cross-section can be transformed from the unobservable variable b to the observable variable, θ. The solid angle element is $d\Omega = d\phi \sin\theta d\theta$.

$$d\sigma = 2\pi b db = 2\pi b(\theta)(db/d\theta)d\theta \sim b(\theta)(db/d\theta)d\Omega/\theta$$
$$d\sigma/d\Omega = (b/\sin\theta)(db/d\theta)$$

(3.9)

In the specific case of Rutherford scattering the force is the Coulomb force with potential energy $U(b)$. Rutherford scattering scales approximately as the inverse fourth power of the scattering angle, favoring small angle scatters. It falls with incident energy as the inverse square, This sketch of the derivation has been NR, but as will be shown later, the electric field magnitude increases as the γ factor, while the time over which the force is large decreases as $1/\gamma$, leaving the momentum transfer a constant at high velocities. This is the origin of the "mip" or the constant energy deposit of a UR minimum ionizing particle. A numerical example for a 1-MeV proton scattering off copper gives a differential cross-section of $(18 \text{ b/sr})/\theta^4$. The differential cross-section, Eq. (3.10), has a factor $(Z\alpha)^2$ due to the coherent sum of quantum amplitudes for the virtual exchange of a photon between the incident proton and the Z protons in the point-like nucleus.

$$\theta_R = a/bT = U(b)/T, \; a = (qe^2 Z)/4\pi\varepsilon_o$$
$$(d\sigma/d\Omega)_R \sim (a/T\theta_R^2)^2 = (qZ\alpha)^2(\hbar c/T\theta_R^2)^2 \tag{3.10}$$

The discussion so far applies to a single interaction with a nucleus of charge Ze. The Rutherford cross-section appears to diverge at small angles. It is cut off, however, when the nuclear charge is screened by the orbiting electrons and by Gauss's law, the total charge inside is zero leading to no scattering. A scale for the screening radius is the size of the atom, taken to be the first Bohr radius, a_o, leading to a minimum scattering angle cutoff of $\theta_{\min}$. The mean squared scattering angle, weighted by the scattering cross-section, is, up to logarithmic factors, proportional to the minimum angle since the factors in the integral are θ^2 weighted by $1/\theta^4$ and integrated over $\theta d\theta$, defined to be $\ln()$ in Eq. (3.11), normalized by dividing by the integral of $1/\theta^4$ times $\theta d\theta$. Throughout this text the logarithmic factors will not be given explicitly, but simply indicated. They are of order one since the logarithms are slowly varying with changes of their argument and are available in the references or with a simple Google search online.

$$\sigma_R \sim \pi(\hbar c/T)^2(qZ\alpha)^2/\theta_{\min}^2$$
$$\langle\theta_R^2\rangle = 2\theta_{\min}^2 \ln() \tag{3.11}$$

There are many small angle scatters off different nuclei which compound when a macroscopic length of material is traversed. There are multiple interactions, N_s, during the full traversal. It is conventional to take the particle direction along the x axis, whereas in this text z is consistently the particle direction and particle charge is qe. Consider traversing a length of material L rather than the characteristics of a single scattering. This is a stochastic process and the root mean square (r.m.s.) values of each single scatter should then be added in quadrature. The average square of the multiple scattering angle after N_s scatters with mean free path for a single scatter Λ is:

$$\langle \theta_{ms}^2 \rangle = N_s \langle \theta_R^2 \rangle, N_s = (N_A \rho \sigma_R dz)/A = dz/\Lambda$$
$$\sim (N_A \rho dz/A)[(a/T)^2 \pi] \ln()$$

(3.12)

By convention the multiple Coulomb scattering is related to the radiation length, X_o, which is appropriate to photon radiation in the field of the nucleus, although there is no obvious connection between the two processes. The radiation length will be discussed later, in Section 3.11 in the exploration of bremsstrahlung. Note that the description has been NR so far, using T, and now shifts to a correct SR description, $T \to p\beta c/2$. Note that the NR limit is $T \to pv/2$ which is correct. This definition for radiation length is in length units. A definition in $X_o \rho$ units also appears in the literature which is useful because the density effect is factored out as was already the case with the nuclear interaction length.

$$\langle \theta_{ms}^2 \rangle = (dz/X_o)(4\pi/\alpha)[m_e c^2/(p\beta c)]^2$$
$$1/X_o = 4\lambda_e^2(Z^2 \alpha^3)(N_A \rho/A) \ln()$$

(3.13)

Defining a multiple scattering energy, ε_s, the mean multiple scattering angle can be thought of in terms of a transverse momentum impulse acquired in traversing a length of material. Since multiple scattering is a stochastic, or random walk, process it has the typical square root dependence on the path length. The energy ε_s should be divided by pc in energy units and by β.

$$\sqrt{\langle\theta_{ms}^2\rangle} = [\varepsilon_s/(p\beta c)\sqrt{dz/X_o}] = \Delta p_\perp(ms)/p$$

$$\varepsilon_s = \sqrt{4\pi/\alpha}(m_e c^2) \sim 21 \text{ MeV}, \qquad (3.14)$$

$$\Delta p_\perp(ms) = (\varepsilon_s/\beta c)\sqrt{dz/X_o}$$

For example, for copper, the radiation length is $X_o = 1.435$ cm. For a 10-cm long block of copper a $100\,\text{GeV}$ proton would receive, on average, a transverse momentum impulse of $55.4\,\text{MeV}$ or an r.m.s. scattering angle of 0.55 mrad. Multiple scattering deflections are much more important for low momentum particles, and examples will be discussed at several points later in the text. At low momentum multiple scattering is likely to dominate the error in the measurement of the position of a charged particle.

3.8. Ionization and Energy Loss — β

Ionization refers to the interaction of an incident charged particle with the electrons in the material which is traversed. The nuclei are irrelevant here since the energy transfer goes as the inverse of the mass, m, of the target. This is clear from everyday experience. Basketballs bouncing off stopped cars do not lose energy which they do when bouncing off other basketballs. The nuclei are compact and cause the few wide-angle Rutherford scatters. The electron targets escape from their atomic binding and are free to propagate in the material and to be observed by appropriately designed detectors. The resulting electron currents are the basis of many detectors and instruments.

The energy loss as a function of the bulk material traversed, is explored first as a single electron scatter by a projectile with charge qe, mass m, and impact parameter b, with $a' = qe^2/(4\pi\varepsilon_o)$. The momentum transfer to a single electron follows from Eq. (3.8). The total energy loss incoherently adds up to the total released ionization energy U_I for Z atomic electrons. As before the treatment is NR and logarithmic factors are simply indicated, but again they are slowly varying factors of order one. The electron wavelength appears since the electrons are the particles which carry off the energy. The energy loss is assumed to be small and the incident particle path is assumed to be undeflected. The Z electrons contribute incoherently.

$$d\varepsilon(b) = (\Delta p_\perp)^2/2m = (a'/b)^2/T, a' = a/Z$$

$$dU_I/dz = [(N_A\rho)/A]Z \int d\varepsilon(b)2\pi b\,db \qquad (3.15)$$

$$= [N_A\rho\alpha^2(Z/A)4\pi q^2][\lambda_e(\hbar c)/\beta^2] \log()$$

Previously, in Figure 3.10, it was simply assumed that the energy loss scaled as $1/\beta^2$, now there is *a posteriori* justification. The energy loss per unit longitudinal distance dz, or per unit material ρdz, is a constant for an ultra-relativistic (UR) particle, and to the extent that Z/A is $= 1/2$ for all elements, such a singly charged particle, $q = 1$, deposits $0.16\,\text{MeV/kg/m}^2$ or $1.6\,\text{MeV/gm/cm}^2$. In many texts the notation dE/dx is used for energy loss and z is a charge. In this text E is electric field, x is a transverse coordinate and qe is the particle charge. For example, in copper a UR singly charged particle loses $1.56\,\text{MeV}$ traversing $10\,\text{cm}$.

$$MIP/(Z/A) \sim 0.32 \text{ MeV}/(\text{kg/m}^2) \qquad (3.16)$$

The ionization of materials by charged particles is a process by which many detectors measure properties of such particles. The minimum ionization and the inverse square behavior on β is very evident in Figure 3.17 as is the anomalous behavior of hydrogen with $Z/A = 1$. At large values of γ the energy loss is reasonably constant, which illustrates the concept of the "mip". The relativistic rise of the ionization energy with momentum is due to the fact that at high energies the electric field of the incoming particle stretches transversely. In that case, the particle may interact with several atoms, while only a single atomic interaction was considered above. Clearly, this rise also depends on the density of the material. A plot of the "mip" for different elements is generated by the script "mip_elements" and is shown in Figure 3.18. Tabulated values for elements appear also in Appendix C. The approximately constant

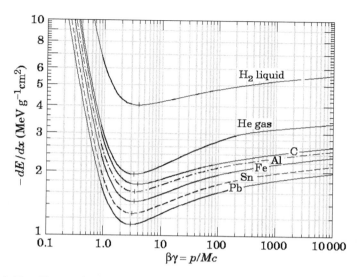

Figure 3.17: Energy loss by ionization in ρdz units as a function of $\beta\gamma$, which is $\sim\beta$ in the NR case and $\sim\gamma$ in the SR case. The "mip" at $1.6\,\text{MeV}/(\text{gm}/\text{cm}^2)$ and the $1/\beta^2$ behavior are very evident.

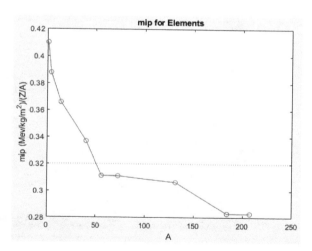

Figure 3.18: Plot of the energy deposit of a "mip" scaled by $1/(Z/A)$ as a function of A. The vertical red line is the approximate constant value quoted in the text. Note the compressed vertical scale.

value of Eq. (3.16) appears as a horizontal line. The concept of a "mip" which is the same for all relativistic singly charged, $q = 1$, particles is a reasonable first approximation to the ionization energy loss.

By measuring the ionization energy deposit in a detector and comparing to the momentum derived using the trajectory in a magnetic field, different mass particles can be identified, as indicated previously. Ionization measurement is fundamentally a measure of the incident particle velocity which works well at low values of β. Data on ionization energy as a function of momentum appears in Figure 3.19 for negative particles produced at the LHC in heavy ion collisions. The separation of the different particle masses from pions to helium nuclei is impressive.

As a charged particle deposits and therefore loses energy, it will ultimately stop in a material which is sufficiently extensive. The particle kinetic energy is now defined to be $T = \varepsilon - \mathrm{mc}^2$. Two simple approximate estimates of the range, or the distance to stop,

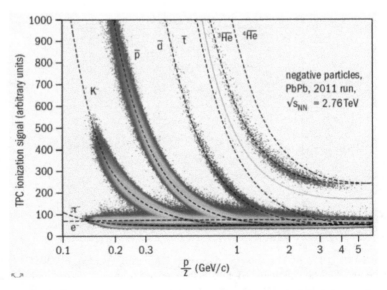

Figure 3.19:   Data from the ALICE experiment at the LHC for the ionization energy deposit of negative particles in Pb–Pb collisions. At low momenta, below about 0.5 GeV, the mass separation of the different particles is clear.

are shown in Eq. (3.17). For a NR particle, the $1/\beta^2$ behavior of $dT/dz \sim 1/T$ means that the range goes as the initial kinetic energy squared or the fourth power of the initial momentum. For a UR particle, if it remains UR over most of its range, $dT/dz \sim$ "mip", so that the range is just proportional to the energy or the momentum. These are useful approximations for making an estimate of the value of the range.

$$\text{Range} \sim T_o^2 \sim p_o^4 (NR)$$
$$\sim T_o \sim p_o (UR) \tag{3.17}$$

In more detail, an incident particle may start out depositing only a "mip". However, as it slows and reaches the end of its range it deposits more energy per unit length, peaking at the end of the range. This is called the "Bragg curve" and a sample appears in Figure 3.20. In fact, this physics phenomenon has recently been adopted in what

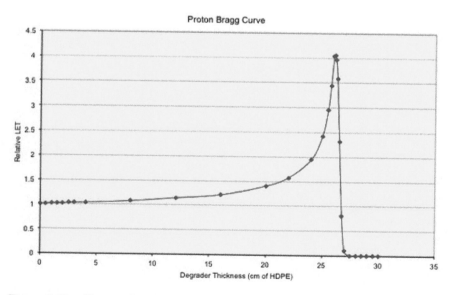

Figure 3.20:   Energy deposit as a function of depth in a material approximating a human body for incident protons. In this case a factor of four reduction of energy deposit in the intervening material is achieved. The patient can also be rotated to further reduce the energy load on any specific intervening tissues.

is called "hadron radiotherapy". It is used at proton or heavy ion accelerators in cancer treatments for deep tumors. The energy of the accelerator is varied so as to deposit the maximum energy at the tumor site, sparing the intervening tissues. This technique improves on X-ray or electron beam radiotherapy. For example, for protons in water, approximating a human body, at $100\,$MeV the range is $7.7\,$cm while at $150\,$MeV it is $16.7\,$cm, and at $200\,$MeV the range is $25.9\,$cm.

3.9. Transition Radiation — γ

TOF, ionization and Cerenkov techniques measure β but their utility collapses at high energy because all β approach one. Transition radiation on the other hand is sensitive to γ which allows the experimenter to identify highly relativistic charged particles. A short preamble on the propagation of photons in a medium with free charged particles, a plasma, gets the discussion started. An electromagnetic wave has a region of frequency where wave propagation is free, bounded by the plasma frequency. Waves below the plasma frequency, ω_p, are reflected. Above it, free propagation is possible. The quantity n_e is the number density of free electrons in the plasma assumed to be fully ionized in Eq. (3.18).

$$\omega_p = c\sqrt{4\pi\alpha n_e \lambdabar_e}$$
$$n_e = N_A \rho (Z/A) \tag{3.18}$$
$$\omega_p/2\pi = f_p = 9.0 n_e^{1/2} \text{ Hz}$$

For example, the atmosphere of the earth is a dilute plasma, the ionosphere, with about 10^{12} e/cm^3. The plasma circular frequency is about 57 MHz, so that AM and shortwave radio "bounce" while FM and TV are strictly line of site. The plasma energy is $\sim 0.91\,$eV times the square root of $\rho(Z/A)$ in MKS units. A material with the density of water and with $Z/A = 1/2$ would have a plasma energy of approximately $20.3\,$eV. For lithium the calculation gives $13.8\,$eV assuming all three electrons are mobile while the measured plasma energy is $6.02\,$eV.

Transition radiation (TR), is emitted when a charged particle crosses a region where the plasma frequency, or the index of refraction, has a discontinuity. It can be thought of as a dipole consisting of the moving charge and the image charge in the medium. The dipole is accelerated and changes direction during the transition and therefore radiates. Note that, compared to the Cerenkov process, the emission for very UR particles extends typically into the X-ray region so that the medium is not fully transparent to its own TR emissions. There are, however, optical photons emitted in the TR process which will be described later in Section 4.21.

It is useful to put the differential cross-section for the emission of the radiation in terms of dimensionless variables. A typical angle for TR is $1/\gamma$ and a typical energy for the emitted photon is γ times the plasma energy which leads to the y and s variables. At each interface pair, or foil, about α such photons are emitted which implies a third scaled variable proportional to the length of the foil, L. In general, scaled variables will always be used when appropriate because they display the underlying physics of the process in question.

$$y = (\theta\gamma)^2, s = (\omega/\gamma\omega_p)$$

$$\alpha \text{ photons, } \varepsilon_\gamma \sim \gamma\hbar\omega_p/3 \tag{3.19}$$

The doubly differential distribution of the energy of the TR photons for a single interface in terms of the scaled variables appears in Eq. (3.20). The factor α/π indicates the electromagnetic coupling factor for the real emission at a single vertex while the factor of ε_γ sets the energy scale for the process. In addition an approximate interference factor for a foil of length L is shown with an interference angle δ and a foil factor $4\sin^2(\delta/2)$. In general interference effects in traversing a foil or a stack of foils will not be covered further.

$$d^2(\hbar\omega)/dsdy = (\alpha/\pi)(\gamma\hbar\omega_P)(y)[1/(1+y) - 1/(1+1/s^2+y)]^2$$

$$\delta = (\omega L/2c\gamma^2)(1+1/s^2+y), 4\sin^2(\delta/2), \tag{3.20}$$

$$U = \hbar\omega$$

The doubly differential cross-section in Eq. (3.20) as a function of y and s is displayed in Figures 3.21 using the Matlab "surf" utility

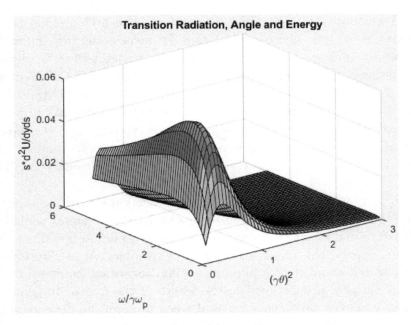

Figure 3.21: Surface of the doubly differential energy radiated in TR as a function of s and y multiplied by s. The peaking at s and $y \sim 1$ is evident, which justifies the use of these scaled variables

with the script "TR_d2N_dtdom". The multiplication by s removes the $1/\omega$ behavior which is characteristic of radiative processes which preferentially emit low energy photons. The peak at $y \sim 1$ is prominent as is the peak at $s \sim 1$.

The distribution in Eq. (3.30) can be integrated over s or y, with results shown in Eq. (3.21) where numerical constants have been dropped in order to focus on the s and y dependence.

$$d(\hbar\omega)/ds = [(1 + 2s^2)\ln(1 + 1/s^2) - 2]$$
$$d(\hbar\omega)/dy = y/(1 + y^2)^2$$
(3.21)

The one-dimensional angular and energy distributions for TR photons are shown graphically in the plots also created by the script "TR_d2N_dtdom" in Figures 3.22 and 3.23. For a single interface at normal incidence the angular distribution of the energy is approximately, $d\omega/d\Omega \sim \beta^2 \sin^2\theta/(1 - \beta\cos\theta)^2$ which is $\sim y/(1 + y^2)^2$ in

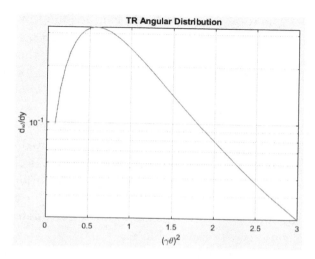

Figure 3.22:   Angular distribution of TR energy U as a function of the variable y. The angular peak occurs at $y \sim 1$.

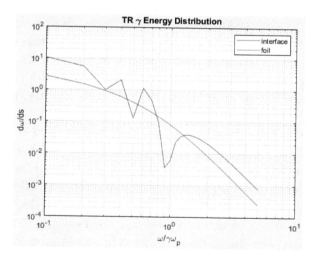

Figure 3.23:   Angular distribution of TR energy U as a function of the variable s. The distribution (red) falls monotonically with photon energy. An approximate interference shape for a foil, of length L, (blue) also appears, Eq. (3.20).

the UR limit. The total energy radiated can be found by integrating again, with results approximately those shown in Eq. (3.19). The number of photons appears to be large because there is no ω dependence in the angular distribution. With a reasonable cutoff for the low energy photons which are absorbed in the material, the number emitted at an interface is $\sim\alpha$.

The angular distribution peaks at $y \sim 1$ or $\theta \sim 1/\gamma$ as expected. However, the photon energy distribution, integrated over angle decreases monotonically with energy. Emission for UR electrons can be in the X-ray region. The effective energy distribution does peak however, because low energy photons are absorbed in the material that forms the interface. Nevertheless, optical TR photons are used in some applications, as discussed later.

For a perfect conductor, with effectively an infinite plasma frequency, there is both forward and backward radiation emitted after traversing a foil. The problem can be treated as the motion of the charge and a dynamic image charge. At normal incidence, $dP/d\Omega_{fb} \sim \beta^2 \sin^2\theta/(1 - \beta^2 \cos^2\theta)^2$ as shown in the output of the App "TR_Polar" in Figure 3.24. At low velocity, the distribution is a dipole, while for UR electrons the dipole morphs into sharp forward/backward behavior. The user should run the App and see how the behavior unfolds using the "Slider" to tune the particle velocity.

In large collider detectors a stack of foils is used to have good detection efficiency by creating a sufficient number of X-ray photons. With only about α photons emitted per foil the number of foils is large. The X-ray photons are detected with gas filled detectors containing a high Z gas such as Xenon using the electrons from the photoelectric effect in detectors as will be discussed later. As also discussed later, optical photons are used in accelerators using optical detectors to explore the properties of UR beams with OTR = Optical Transition Radiation.

Data from the ATLAS experiment at the CERN Large Hadron Collider (LHC) is shown in Figure 3.25. The discrimination between pions, of mass 139.6 MeV and electrons, mass 0.511 MeV is provided by the existence of keV X-rays emitted by the electrons. In the

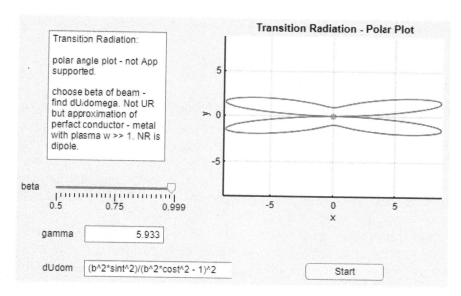

Figure 3.24: Angular distribution for TR in a metal. The initial value of β is chosen by "Slider".

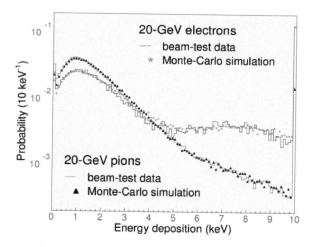

Figure 3.25: Test beam data of a TR detector used in the ATLAS LHC experiment. The electron emission of X-rays with energies >5 keV provides particle identification (PI) for electrons.

20 GeV beam, the pions have a γ factor of 144 while the electron factor is 39,139.

3.10. Compton Scattering — γ

Compton scattering is a process wherein an incident photon is scattered by a charged particle, initially at rest. The process at the lowest photon energy is called Thomson scattering, as was discussed in Section 2.2 with a cross-section of $\sigma_T = (8\pi/3)r_e^2$. In this case, the photon energy was unchanged by the scattering. To explore high energy scattering it is necessary to first explore the full UR kinematics as introduced previously in Section 2.7. The conserved quantities are energy, $\varepsilon = \gamma mc^2$, and vector momentum $p = \gamma\beta mc$. Often units of energy are used for energy, momentum and mass by simply setting $c = 1$ and adding c factors back as desired for numerical calculations. However, in this text all the factors of c will normally be used to keep the dimensions clear.

The Compton cross-section, $(\gamma - e)$, compared to the NR Thomson cross-section, σ_T, is approximately, up to logarithmic factors, given by the Klein–Nishina cross-section, σ_{KN}, shown in Eq. (3.22). The photon energy is reduced by the scattering in contrast to Thomson scattering. The quantity s is the total center of mass (CM), energy squared of the photon plus electron system. Numerically, the cross-section, without the logarithmic factor, is $\sim 0.25\ b$ when the CM energy is $m_e c^2$, falling at high energies approximately as 87 $nb/[s(\text{GeV}^2)]$. The existence of Compton scattering showed that light could act as a particle and not a wave leading to the award of the Nobel prize in 1927.

$$\sigma_{KN} \sim (3/8)\sigma_T(m_e c^2/\sqrt{s})^2[1 + \ln()] \tag{3.22}$$

The Compton wavelength, λ_e, is $h/(m_e c)$. In the text, the reduced Compton wavelength has been consistently used. Conservation of energy and momentum leads to the famous Compton relationship for the wavelength shift of the scattered photon and the electron Compton wavelength, $\lambda - \lambda_o = \lambda_e(1 - \cos\theta)$. This shift occurs because the photon loses energy by scattering. The angle θ is the outgoing photon angle, while ϕ is the final state electron angle in Eq. (3.23). Dimensionless variables s and y are used in the kinematic expressions as was the case for transition radiation. The variable s is the ratio of the outgoing to incoming photon energy, while y is the ratio of incoming photon energy to the electron rest energy. Energy and momentum conservation give three equations (planar topology). They can be used to remove reference to the outgoing electron, yielding $\omega_o \omega(1 - \cos\theta) = mc^2(\varepsilon - mc^2)$ which is equivalent to the first line of Eq. (3.23) since energy conservation for the electron is $\varepsilon - mc^2 = \hbar(\omega_o - \omega)$.

$$\omega/\omega_o = 1/[1 + y(1 - \cos\theta)] = s, y = \hbar\omega_o/m_e c^2$$

$$\tan\phi = 1/[(1 + y)\tan(\theta/2)] \tag{3.23}$$

$$\hbar(\omega_o - \omega)/m_e c^2 = 2/[-1 + (1 + 1/y)^2/\cos^2\phi]$$

The photon can always be backscattered as expected from the discussion of Section 2.7. The kinematics specific to Compton scattering are worked out in the App "Compton_Polar". The user can choose the incident photon energy via the "Slider". The backscattered energy is output via an "EditField". The backscattered energy depends only on the incident photon energy and the electron mass, making a measure of that energy a measure of the beam energy. Plots shown in Figure 3.26 are made of the scattered photon angle contour and for the recoil electron angle as a function of the photon angle. The backward value of s, is $s_b = 1/(1 + 2y)$.

This two-body elastic process is completely defined kinematically by the incoming energy and the scattering angle. The dynamics define the single differential cross-sections for photon energy and angle which are expressed in terms of the dimensionless s and y variables in Eq. (3.24). The total cross-section as a function of initial photon energy is also shown below. Only the functional dependence is shown in Eq. (3.24) except for $d\sigma/d\Omega$ which has all the numerical factors and is proportional to the low energy Thomson cross-section

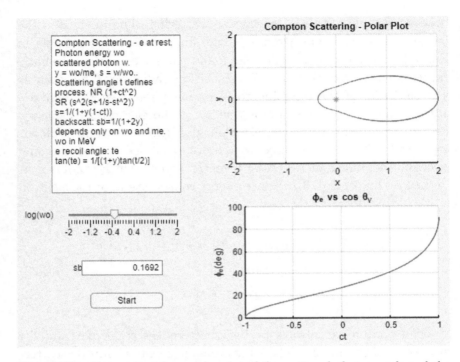

Figure 3.26:   App output for a polar plot of the scattered photon angle and the electron recoil angle as a function of the cosine of the photon angle. The incident photon energy is in MeV units.

defined previously in Eq. (2.6).

$$d\sigma/d\Omega = [(\alpha \lambda_e)^2/2]s^2[s + 1/s - \sin^2\theta]$$
$$d\sigma/d(\hbar\omega) = (1/y^2)[s + 1/s - (2/y)(1 - s)/s + (1/y^2)(1 - s)^2/s^2]$$
$$\sigma(\hbar\omega_o) = [(y^2 - 2y - 2)\ln(1 + 9y^2 + 8y + 2)]/(4y^4 + 4y^3 + y^2)$$

$$(3.24)$$

The App "Compton2" provides two sets of three plots. The initial photon energy can be varied using the "Slider". The plots are for those quantities shown in Eq. (3.24) augmented by other plots. The set of plots is chosen using the "Button" utility to choose the menu here. In Figure 3.27, the plots are kinematic; photon energy and

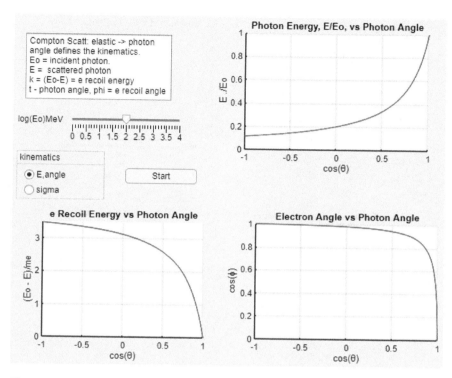

Figure 3.27: Kinematic plots for Compton scattering as a function of photon angle.

angle and electron angle are all plotted as a function of the photon angle. The backscattered photon is quite evident. In Figure 3.28, the plots reflect the dynamic of the scattering process. The three plots are the angular distribution the photon energy distribution and the total photon cross-section as defined in Eq. (3.24). As usual, the user is encouraged to see how the dynamics changes with the incoming photon energy.

This has been a lengthy section covering aspects of SR kinematics and dynamics. In future discussions, the exposition will be much more cursory. For example, the SR aspects of the next section will be presented much more sketchily.

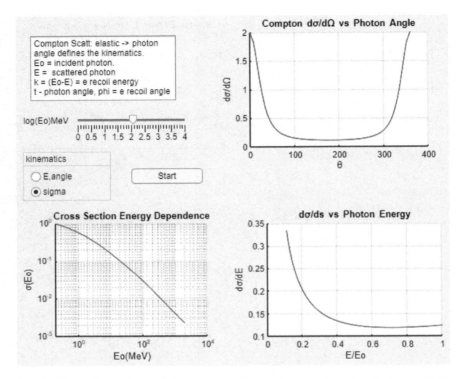

Figure 3.28: Second set of plots generated by "Compton2". The photon angle is given in degrees. The shapes are set by the dynamics.

3.11. Bremsstrahlung and Pair Production — γ

Now consideration moves to the radiation of photons due to the acceleration of charged particles. Since electrons are the lightest such charged particles, it is to be expected that the treatment will largely apply to electrons. First it is useful to consider two aspects of cross-sections, vertex counting and coherence. Vertex counting is simply asking how many electron–photon vertices exist for a given process. The quantum amplitude for any process gains a factor qe for each vertex, and the cross-section is proportional to the square of the amplitude, so that each vertex contributes a factor α to the cross-section. A previously discussed process is Compton scattering which has two vertices, and thus a cross-section factor of α^2. Both bremsstrahlung and pair production by photons have an additional vertex associated with the emission of a new particle and therefore scale as α^3. In order to count the number of vertices one needs either to draw the appropriate Feynman diagram or just accept the scaling power of α which is quoted.

In quantum mechanics the total amplitude for a process is the sum of the individual amplitudes. The cross-section is proportional to the square of the total amplitude. If there are Z processes and if they are coherent, or in phase, the cross-section is Z^2 times the individual cross-section. If not, the sum over processes is incoherent and the total cross-section is just Z times the individual cross-section. Bremsstrahlung and pair production are both coherent processes over the Z protons in the nucleus. Compton scattering is incoherent over the atomic electrons. Very approximately the difference can be understood because the atomic size is ~ 0.1 nm while the nuclear size is $\sim 1 fm$, Eq. (2.1). In NR quantum mechanics (QM) the momentum, $p = mc\beta$ is equal to $(h/2\pi)k$ for a plane wave with phase kz. The phase is then $mc^2\beta L/(h/2\pi)c$ for a system of size L. For electrons with $L \sim a_o$ in atoms the phase is $\sim 135\beta$ which is large so the process in incoherent. For nuclear sizes, $\sim 10^5$ smaller, the process is coherent.

In the NR case, the simplest radiation is caused by an oscillating dipole, moment p_o, with an oscillation frequency ω. From a dimensional point of view the static dipole field can be "transformed" by the trick of multiplying by the dimensionless factor, (ar/c^2) where the acceleration a is $\sim p_o\omega^2$. The field then goes as $1/r$ and the energy density in the field goes as the electric field squared, $u = \varepsilon_o E^2/2$. This is a radiation field that propagates a time averaged power, $\langle P \rangle$, proportional to $cr^2 E^2 \varepsilon_o$, which is independent of r. The angle θ is that between the dipole direction and the observer. The angular distribution of the radiated power and the average power are displayed in Eq. (3.25). If the radiated power is treated as the acceleration of a particle of charge qe rather than an oscillating dipole moment, the NR power depends on the square of the acceleration and the fine structure constant, α.

$$d\langle P\rangle/d\Omega \sim \omega^4 p_o^2 \sin^2\theta/(8\pi\varepsilon_o c^3) = (\alpha\hbar c/4\pi c^3)(qa)^2 \sin^2\theta$$

$$\langle P\rangle = \omega^4 p_o^2/(12\pi\varepsilon_o c^3) \sim (qea)^2/(6\pi\varepsilon_o c^3) = (qa)^2(2/3)(\alpha\hbar c)/c^3$$

$$(3.25)$$

Bremsstrahlung is the emission of radiation by an accelerated projectile of mass m_{pr}, of charge qe, in the Coulomb field of the nucleus, of charge Ze. The photon spectrum follows a $1/\omega$ behavior, has $Z^2\alpha^3$ behavior and is proportional to the projectile reduced Compton wavelength squared. The $1/\omega$ behavior in NR quantum mechanics follows from the fact that the amplitude in second-order perturbation theory has a factor which is the inverse of the energy difference of the initial and final states, which is the photon energy. This type of behavior was previously seen in the discussion of transition radiation and is quite general for radiative processes. A schematic view of the process is shown in Figure 3.29.

In the relativistic case, the angular distribution for radiation is thrown more forward, which is called the searchlight effect, a common effect of SR. Assuming velocity and acceleration are parallel the NR

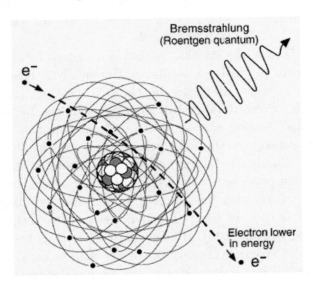

Figure 3.29:   Very schematic view of the bremsstrahlung process. Note that, if drawn to scale the atomic electrons would have a size 10^5 times that of the nucleus.

dipole angular distribution shown in Eq. (3.25) acquires a factor $1/(1 - \beta \cos \theta)^5$. This type of behavior will be explored again in the discussion of synchrotron radiation in Section 4.20.

$$d\langle P\rangle/d\Omega = (\alpha\hbar c/4\pi c^3)(qa)^2 \sin^2 \theta/(1 - \beta \cos \theta)^5 \qquad (3.26)$$

Converting from power to the differential cross-section as was done for Thomson scattering in Eq. (2.6), results in the bremsstrahlung differential cross-section shown in Eq. (3.27). A scaled variable, y, the radiated photon energy divided by the incoming projectile energy, ε_o, is used to factor out the relevant dimensional variables. The factor $Z^2\alpha^3$ expected from vertex counting and coherence arguments appears as does the Compton wavelength of the projectile since it is the projectile that is accelerated and radiates. The $1/\omega$ behavior is also now familiar from NR radiative processes. In this specific case, the log() factor which is normally ignored is shown explicitly but is only an approximation to the full expression. It is here that the radiation length appears naturally. However, recall that it is the projectile radiation length not the normally tabulated electron radiation length.

$$d\sigma_B/dy \sim 4Z^2\alpha^3\lambda_{pr}^2(1/y)[4/3(1 - y) + y^2]([\ln(184/Z^{1/3})])$$
$$y = \hbar\omega/\varepsilon_o \qquad (3.27)$$
$$d\sigma_B/dy = (A/N_A\rho X_o)(1/y)[4/3)(1 - y) + y^2]$$

Integrating over all photon frequencies, a constant cross-section is found. The radiation length is defined to be the mean free path for bremsstrahlung in macroscopic material. For example, in lead, the radiation length is 0.56 cm.

$$\sigma_B = \int_0^1 (d\sigma_B/dy)dy = A/(N_A\rho X_o)$$
$$X_o^{-1} = 4(N_A\rho/A)(Z^2\alpha)(\alpha\lambda_{pr})^2 \ln(184/Z^{1/3}) \sim Z^2\alpha^3/m_{pr}^2 \qquad (3.28)$$

The energy loss due to bremsstrahlung is $d\varepsilon/dz = -\varepsilon/X_o$. Numerically, the radiation length is very approximately 1800 $(A/Z)/Z$ (kg/m^2) for electron projectiles. A plot of the tabulated radiation length in $X_o\rho$ units vs. A/Z^2 appears in Figure 3.30 as

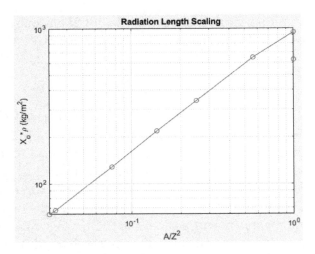

Figure 3.30:   Plot of radiation length with the density factored out as a function of A/Z^2. The log–log plot indicates that a power law fit, almost linear, would give an adequate first approximation to the true results if very low Z values were excluded.

generated using the script "Xo_A_Z_Scaling". The two points on the right are for H and He, both with the same value of A/Z^2. Tabulated results for radiation lengths of elements appear in Appendix C. Electrons in a beam can be identified by detecting photons emitted when they are bent in a beam line steering dipole in a non-destructive fashion. Alternatively, they can be identified destructively by interposing a few radiation lengths of material in the beam and observing the subsequent large energy deposition due to radiation in addition to normal ionization.

Pair production by photons is a process very closely related to bremsstrahlung. It can be thought of as the "decay" of a photon into an electron–positron pair, which is allowed in the field of a nucleus because the intermediate electron is "virtual" and need not have the mass of a free electron. A photon cannot "decay" in vacuum because energy–momentum cannot be conserved. A strong field is required which means that the photon interacts with the protons in the nucleus, not the electrons. A schematic view of the process appears in Figure 3.31. The cross-section for an incident photon is expected

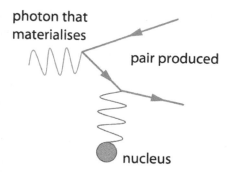

Figure 3.31: Schematic view of photon pair production using the strong electric field of a heavy nucleus.

to scale as $Z^2\alpha^3$ since there are again three electromagnetic vertices, one within the nucleus, and the amplitudes are again expected to add coherently.

The electron can have any energy ε greater than the rest energy, up to the photon energy, ε_o less the rest energy which are the limits given for s below in Eq. (3.29). Integrating over the electron energy, the total pair production cross-section is approximately a constant. Typical electron angles are $\sim m_e c^2/\varepsilon_o$. It should be mentioned that there is a higher order, weaker, process related to bremsstrahlung. That occurs when the photon emitted by the projectile has sufficient energy to virtually create an electron–positron pair. There is a fourth vertex, so the process is typically a factor α weaker than bremsstrahlung. Scaled variables are again used for the electron energy divided by the incident photon energy and the incident photon energy scaled to the electron rest mass.

$$a_1 = [s^2 + (1 - s)^2 + (2/3)s(1 - s)]$$
$$a_2 = a_1 [\ln(2ys(1 - s)) - 1/2]$$
$$d\sigma_p/ds = a_2 [\ln(184/Z^{1/3}) - s(1 - s)/9] \quad (3.29)$$
$$s = \varepsilon/\varepsilon_o$$
$$1/y < s < 1 - 1/y, y = \varepsilon_o/m_e c^2$$

The nuclear charge Z can be screened by the atomic electrons by an amount which depends on the impact parameter of the photon. The basic cross-section, with a factor $Z^2\alpha^3\lambda_{pr}^2$, which is suppressed for clarity, is given by a_1 with a logarithmic factor given explicitly as a_2. The effect of screening is approximately given as a third factor and is explicitly Z dependent. The unscreened result is $d\sigma_p/ds \sim a_2$. The total pair production cross-section compared to the closely related bremsstrahlung cross-section is $\sigma_p = (7/9)\sigma_B$. The two processes are displayed in some of their aspects in the App "Pair_Brem". A specific output is shown in Figure 3.32. The incoming energy, electron or photon, is chosen by "Slider" as is the Z of the nucleus. The top plot is the electron spectrum for pair production.

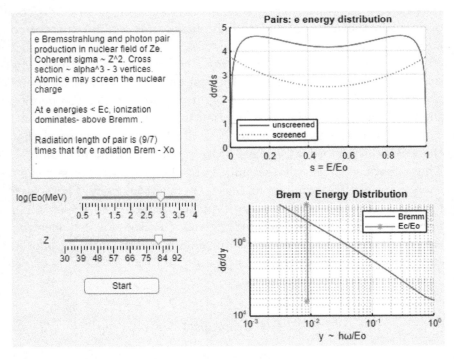

Figure 3.32: Output of the App "Pair_Brem". For pairs the electron energy distribution is shown, while for bremsstrahlung the photon spectrum is displayed. The critical energy when bremsstrahlung energy loss equals ionization energy loss, Eq. (3.31), is also indicated. Scaled variables are used.

If the photon could decay with an isotropic distribution in angle, the energy spectrum would be flat which is approximately valid here. The effects of screening for different Z can be explored. In the bremsstrahlung case, the typical radiative fall of photon cross-section with photon energy is seen. The vertical line shows the "critical energy", discussed in the following section, below which the energy loss of an electron is dominated by ionization and not bremsstrahlung radiation.

Comparing the strengths of the nuclear and electromagnetic processes, coherence plays an important role in making the strong interactions less important than the intrinsically weaker electromagnetic ones. This is important in particle identification, ID, since electrons and photons interact rapidly in materials with a large Z, whereas hadrons, for example protons, do not. Over most of the periodic table the electromagnetic cross-sections actually exceed the hadronic cross-sections, even though the basic electromagnetic coupling is weaker than the hadronic coupling defined by the "strong" interactions. The strong cross-sections, σ_h, appear in Eq. (2.7) while the bremsstrahlung cross-section, σ_B, appears in Eq. (3.28) for electron projectiles. The ratio scales approximately as $Z^2/A^{2/3}$, with the factor of 5 inserted to make the agreement good at high values of Z as shown in Eq. (3.30).

$$\sigma_h \sim A^{2/3}\lambda_p^2, \quad \sigma_\gamma \sim (Z\alpha)^2 \alpha \lambda_e^2$$
$$\Lambda_I/X_o \sim [Z^2/(5A^{2/3})] \tag{3.30}$$

For example, in beryllium, the radiation length is approximately equal to the nuclear interaction length and is less than it for heavier elements using the tabulated quantities from Appendix C. This effect is the basis of destructive particle identification where a beam particle interacts and the characteristic length in heavy material is, if short, taken to be an electron or photon. A plot of X_o and Λ_I for a few representative elements appears in Figure 3.33. Examination of interactions in lead, for example, is used to distinguish clearly between electrons and "strongly" interacting particles, called "hadrons" such as pions, kaons or protons.

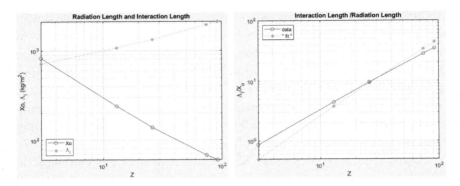

Figure 3.33: Plot of the radiation length and nuclear interaction length as a function of Z for a few representative elements. The left plot is for the X_o and Λ_I separately and the right plot is for the ratio. The "fit", is Eq. (3.30), with $Z^2/A^{2/3}$ scaling. The density effect has been scaled out by multiplying X_o and Λ_I by the density.

3.12. Critical Energy and Muons — γ, β

The energy loss of a particle due to ionization falls as $1/\beta^2$ while that for bremsstrahlung increases with the projectile energy, $d\varepsilon = -\varepsilon dz/X_o$. At some point bremsstrahlung will dominate. The energy at which this occurs is defined to be the critical energy. Using Eq. (3.15) for ionization and Eq. (3.28) for bremsstrahlung, ignoring the two different log() factors for the two processes and assuming a "mip" with $\beta = 1$, the critical energy for the projectile is shown in Eq. (3.31). Using Eq. (3.31), the estimated critical energy is 221 MeV/Z for electron projectiles. The actual critical energy for electrons is approximately fit to the data in solids and gives electron critical energies of order 10 MeV for heavy elements, for example 7.33 MeV for lead.

$$\varepsilon_c \Rightarrow \varepsilon_B/\varepsilon_I = 1$$
$$1 = [Z^2\alpha(\alpha\lambda_{pr})^2\varepsilon_c]/[\pi q^2 Z(\alpha^2\lambda_e\hbar c)]$$
$$\varepsilon_c \sim (\pi q^2)(\lambda_e\hbar c/\lambda_{pr}^2)/(Z\alpha) = \pi\hbar c/[\lambda_e\alpha Z] \qquad (3.31)$$
$$(\varepsilon_c)_e \sim 610 \text{ MeV}/(Z + 1.24)$$

The critical energy scales as the square of the projectile mass since heavier projectiles are not accelerated as much as light ones.

The forces on electrons and muons are the same but radiation scales as the square of the acceleration not the force. Indeed, muons, mass 105.7 MeV compared to electrons with mass 0.511 MeV, do not radiate appreciably at "low" energies but simply lose energy by ionization. For example, the muon critical energy in iron is about 400 GeV, which reflects the mass ratio squared of muons to electrons of about 4.27×10^4. An approximate fit to the more exact calculation of the critical energy for muons is

$$(\varepsilon_c)_\mu \sim 7980 \, \text{GeV} / (Z + 2.03)^{0.88} \tag{3.32}$$

Muons can therefore be identified as those charged particles that do not interact strongly and which also do not radiate appreciably at energies less than a few 100 GeV. A beam of muons can be made by first creating secondary pions in proton interactions with a target of thick material. The emerging pions and other strongly interacting particles can be removed by stopping them in a "beam dump" of thick and heavy material. What remains are the muons and neutrinos from the decays of pions, $\pi \rightarrow \mu + \nu$, that occur before being absorbed. The muons simply lose energy by ionization in the absorbing material. In fact, muons are often used to calibrate detectors because they only deposit a fixed amount of ionization energy, a mip, into any sensitive detector independent of momentum to first order.

A simple model of a muon beam propagating in material is explored in the script "MuonBeam_App". The muon beam suffers only energy loss and multiple scattering, without bremsstrahlung or other interactions. A modeling technique called Monte Carlo uses the Matlab utility "rand" to generate distributions of kinematic quantities. The function "Gaus" picks out of a Gaussian distribution. The function "Mult" populates multiple scattering and "Eul" uses Euler angles to go from a coordinate system with z the muon direction to the laboratory frame where z is the beam line axis.

"Sliders" are used to pick the initial beam kinetic energy T_o and the material length L. The initial beam spread in x and y and the number of strips of L are fixed as is the initial beam momentum spread, dp/p. A "DropDown" menu is used to pick the element making up the material. A "movie" is made of the first 5 muons as they are tracked through the strips. Overall 1000 muons are tracked. The "EditField" tool is used to record p_o, $(d\varepsilon/dz)L$, the approximate range, and the approximate multiple scattering angle, scaling as $\sqrt{L}$. The actual mean absolute values of the final energy and x are displayed along with the number which stop in traversing L. An example with many stopping muons is shown in Figure 3.34. Note

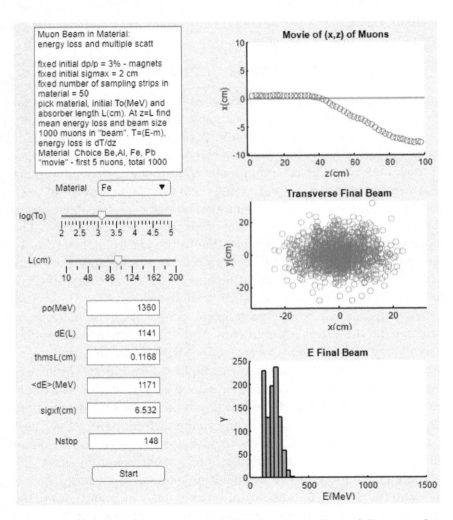

Figure 3.34: A muon beam example which is near to being fully stopped in 100 cm of Fe. Approximately 1.2 GeV of energy is lost and exit angles of 117 mrad typically occur. The beam size has gone from 2 cm to 6.5 cm and ∼15% of the muons have stopped.

that the path length is greater than L due to the scattering but is distributed. The user can choose "Slider" values to explore a wide range of possible beam configurations.

3.13. Neutrino Beams and Muons

What about the neutrinos that accompany the muons when the pions decay? A schematic of a typical neutrino beam appears in Figure 3.35. Protons strike a thick target and charged secondary particles, mostly pions, are focused by a magnetic "horn". The horn has axial currents flowing in opposite directions at small and large angles. Pions with smaller angles than the inner current are undeflected. Pions ate larger angles encounter an azimuthal magnetic field which focuses the pions of a chosen charge, positive or negative, to be brought parallel to the beam axis in both x and y.

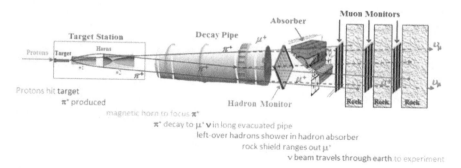

Figure 3.35:   Schematic diagram of a neutrino beam showing target, horn, decay vacuum pipe, and absorber sequence.

A large diameter vacuum pipe follows where pions (and some kaons) decay into muons and neutrinos. The beam still contains hadrons, which are removed by interposing several collision lengths of dense material. The neutrinos proceed to a distant experiment while the muons are monitored in a series of detectors which will be discussed later. The accurate measurement of the muon flux allows for a measure of the neutrino flux and a prediction of the neutrino flux which remains after all the muons have been absorbed by ionization loss and brought to the end of their range.

As has been mentioned previously, the muon critical energy is of order one TeV, while the electron critical energy is a few MeV. A circular electron accelerator at high energy radiates a significant amount of energy in the form of photons as will be quantified in the later discussion of synchrotron radiation. That energy loss, which scales as γ^4, then limits the affordable energy of an electron circular storage ring due to the very large wall power needed to operate the ring for experiments.

In contrast, muons do not have a problem of radiation, but they do have a problem in that they have substantial transverse momenta due to the parent pion transverse momentum and the momentum they receive from the pion to muon decay process. In addition, they decay with a lifetime at rest of 2.2 μs. A possible "muon collider" needs to very rapidly "cool" the muons so they fit into the ring emittance. One possible approach is called "ionization cooling". The muons first transverse some material. This reduces both the transverse and longitudinal momentum components by means of ionization energy loss. The muon is then accelerated, increasing the p_z momentum, thus reducing the dx/ds transverse phase space, and achieving a net cooling. No muon collider has yet been built, but the promise of a low power and a high energy operation makes it an attractive possibility. An R&D experiment has recently established experimentally that the ionization cooling principal is valid. Parenthetically, an analogous situation occurs during acceleration, when the transverse phase space of the accelerated beam is reduced. The use of invariant phase space is defined later in Chapter 4.

3.14. Electromagnetic Cascade — X_o, Λ_I

Electrons radiate bremsstrahlung photons in matter. In turn, those photons can make electron–positron pairs. The result is that an electron makes a cascade, or "shower" in high Z material. The number of particles in the shower rises geometrically until the critical energy is reached after which new particle production ceases and the charged particles stop by ionization loss (range out) while the photons are ultimately absorbed by the photoelectric process.

Understanding a shower uses many of the topics covered already; ionization, bremsstrahlung, multiple scattering, critical energy, photoelectric effect and pair production.

The very simplest cascade model is that there is a geometric rise in the number of particles in a "shower" since each interaction occurs, on average, in one X_o and there are two particles per interaction. That implies that the number of particles in the shower at depth t is $N(t) = 2^t$, where $t = z/X_o$. The produced particles are assumed to share the initial energy, ε_o, equally $-\varepsilon(t) = \varepsilon_o/N(t)$. The growth of the shower continues until the particles in the shower reach the critical energy and multiplication ceases. After that shower depth the electrons and photons are removed by ionization loss (electrons) or photoelectric absorption (photons). This change occurs at "shower maximum" $t_{\max} = \ln(\varepsilon_o/\varepsilon_c)/\log(2)$ with $N_{\max} = N(t_{\max})$.

At the "shower maximum" all particles have the critical energy and a typical transverse momentum due to the production process is set by the electron mass. It is notable that a shower can be contained in a length of material which grows only slowly with energy, which makes shower detectors quite compact. It is also notable that the energy can be estimated simply by counting the number of particles in the shower. That fact is the basis of energy measurements of electrons and photons, called calorimetry, with an example appearing in Section 3.20. A typical transverse shower size is set by the larger multiple scattering energy ε_s and is defined by the Molière radius, r_M. For lead Eq. (3.33) gives $r_M = 1.56$ cm. Another fit is $r_M = 0.0265X_o(Z + 1.2)$ which gives 1.24 cm for lead.

$$\langle p_\perp \rangle \sim m_e c, \varepsilon \sim \varepsilon_c$$
$$\langle \theta \rangle_{prod} \sim m_e c^2/\varepsilon_c$$
$$\langle \theta \rangle_{ms} \sim \varepsilon_s/\varepsilon_c > \langle \theta \rangle_{prod} \tag{3.33}$$
$$r_M \sim \varepsilon_s X_o/\varepsilon_c, \ r_M \rho \sim (70 \text{ kg/m}^2)(Z/A)$$

Going a step beyond this level of approximation a very simple Monte Carlo modeling of a shower is made based on the Matlab random number generator "rand". A solid block of material, lead, $30X_o$ long is split into 60 strips. The initial electron energy is chosen by "Slider". The electron, if it interacts, using exponential weighting with length scale X_o, in a strip makes a photon with energy scaling as $1/\omega$ and loses energy ω. If it does not it simply deposits ionization energy of magnitude the strip "mip" divided by β^2 and also multiple scatters in the strip. Any photon in the shower may interact and create an electron–positron pair with an exponential weight in z of the pair production mean free path, $9X_o/7$.

The script "Electron_Pb_Shower" then tracks the shower through all the strips. The transverse dimension is only explored very approximately. The App has two subsidiary functions "Elec" and "Gam" to track an electron/positron or a photon through a strip. Information is shared using the "global" statement. The main function is "update" which steps through the strips and keeps track of the energy deposited in a strip, the number of particles, their type (electron or photon) and their approximate transverse size. Specific results are shown in Figure 3.36. The exact results are very sensitive to a cutoff energy of the particles in the shower. The distances are increased by multiple scattering which is only approximately tracked. Nevertheless, for initial energies between 100 and 2000 MeV the results, seen in Figure 3.36, are an improvement on the simplest model. Plots are made using the Matlab utilities "bar" and "histogram".

In Figure 3.36, the top plot is the energy estimated to be deposited in each strip, starting with just the ionization energy of the incident electron until it interacts. The number of shower particles above an energy cutoff appears in the lower plot, where a stack of electrons/positrons is in blue, with photons in red. There are no photons until the initial electron interacts. The shower starts to die off slowly after shower maximum. A second set of two plots can be invoked using the "Button". The second set of plots appears in Figure 3.37. The top plot shows the energies of any particles remaining at the end of the material. Although they are "active" they are kept only because they are slightly above a low cutoff energy.

In fact, the true representation of a shower is much more complex. The scattering angles become large near the end of the shower deposits and the physics is much more complex than is attempted

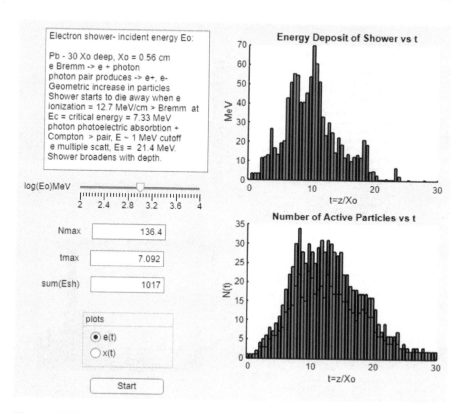

Figure 3.36:   Results of the choice of a 1-GeV electron shower for a specific
Monte Carlo run. The maximum multiplicity and depth for "shower maximum"
as defined in the simple model are shown. The total shower energy deposit above
a low cutoff is shown as sum(E_{sh}) in an "EditPlot" for this Monte Carlo run.
Electrons are blue, while photons are red.

here, where only the simplest model, appropriate for the early part
of the shower is worked out. Even so, there are significant variations
in the shower development from electron to electron for this simple
model, starting from the variations in the initial electron interaction
point. Multiple runs give an idea of shower to shower variations.

A much more detailed Monte Carlo model of an electromagnetic
cascade is shown in Figure 3.38. Standard programs to follow an
electromagnetic shower exist and have been tested in great detail.
Fortunately, the physics is well understood at all energy scales for

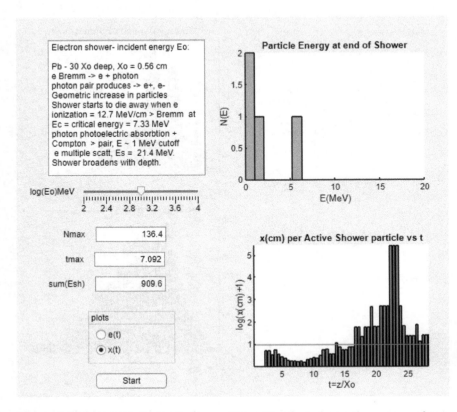

Figure 3.37:   A second set of shower plots. The first shows the energy of any "active" particles at the end of the shower. The second plot shows an approximate transverse shower size. The size grows with shower depth as the shower particle energy decreases. The red line is the Molière radius, appropriate as a typical size at shower maximum.

photons and electrons. There are many low momentum wide angle tracks in this much more complete model, a model which is necessary to accurately track the shower. As seen a shower is a one-dimensional object only approximately and only early on in the development of the shower. The user can explore these more correct models as desired.

It is clear that studying a shower in detail requires detector resolutions in all three dimensions with pixels of a characteristic

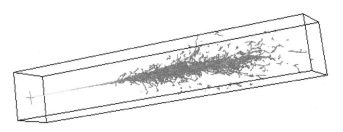

Figure 3.38:   The results of a detailed Monte Carlo modeling of a photon shower. The photon makes an electron/positron pair with very small angles, characteristic of the production mechanism, $\sim m_e c^2/\varepsilon_0$. Subsequent interactions spread the shower transversely until at the end of the shower it is almost isotropic.

size of a radiation length or less. High speed operation is also a priority in many applications. Indeed, recently very high granularity electromagnetic calorimetry has been adopted in experiments based on silicon detectors which offer both speed and spatial granularity.

3.15. Drift and Diffusion — x, y

An overall goal of instrumentation for beams and detectors is to measure the position, momentum and mass (energy) of each beam particle. Achieving that goal may require an ensemble of detectors making different measurements. A charged or neutral particle ionizes the material of a detector as has already been discussed. How is that converted into a detectable signal that allows the determination of the position of a particle? The momentum can then be determined by multiple measurements of the trajectory of a charged particle in a magnetic field. The mass can be determined by deploying differential Cerenkov counters, for example, at least at low momenta.

First, the ionization charge must be detected and made into a signal, for example in a gas. Initially assume there is a uniform electric field, E exerting a force $= qeE$. The typical thermal velocity, v_T, of N_2 at STP is ~ 1 km/s. Electrons are faster with an inverse square root scaling of the velocity in mass as seen in Eq. (3.34) with

~120 km/s at STP. Thermal collisions create an effective velocity dependent opposing force $= Kv$. The acceleration is zero at a velocity called the drift velocity when $v = v_d = (eE)/K$. The mean time between thermal collisions is $\tau = \Lambda/\langle v_T \rangle$. The mean free path is $\Lambda = A/(N_A \rho \sigma)$. The drift velocity is $v_d = (qeE/m)\tau$. The average thermal velocity squared follows from the thermal equi-partition of energy and is $\sqrt{3kT/m}$. It is assumed that the drift velocity is a small perturbation on the thermal velocity. The mobility, μ, is the drift velocity per electric field with the density defined at STP in order to account for the density effect on mobility.

$$\tau = \Lambda/\langle v_T \rangle, v_d = (qeE/m)\tau$$

$$\Lambda = A/(N_A \sigma \rho), \langle v_T \rangle = \sqrt{3kT/m} \qquad (3.34)$$

$$\mu = v_d/[E/(\rho/\rho_{STP})] = qeA/[N_A \sigma \rho \sqrt{3kTm}]$$

There is also a spatial spreading of the ionization as it propagates in the gas or liquid. That effect is due to scattering of the electrons with the material along its path. The diffusion of the ionization is described by the diffusion equation in one dimension shown in Eq. (3.35) where D is the diffusion coefficient. The solution for ρ is a Gaussian in z with an r.m.s. σ_D as can easily be verified by substitution into the second order PDE. The dimension of D is length^2/time. There is a square root dependence of the r.m.s. diffusion distance on the path length. A typical value for gases at STP is $D \sim 0.2\,\mathrm{cm}^2/\mathrm{s}$. For a drift distance of 2 cm with a drift velocity of 10 cm/μs, the diffusion distance σ_D is only 2.8 μm. The diffusion effect, both transverse and longitudinal to the drift velocity, sets a limit to the spatial accuracy with which the ionization can be measured after drifting to a distant detector element. The expression in Eq. (3.35) using temperature is called the "thermal limit" and illustrates the competition for diffusion effects between the thermal random motion and the motion due to the directed electric field. The drift and diffusion processes are complex and the estimates given in Eqs. (3.34) and (3.35) are at best order of magnitude estimates. Recourse to data or a much more detailed analysis is mandatory in this case.

$$\partial\rho/\partial t = -D\partial^2\rho/\partial^2 z$$

$$\sigma_D = \sqrt{2Dt} = \sqrt{2Dz/v_d} = \sqrt{2kTz/(qeE)} \qquad (3.35)$$

Drift and diffusion are explored using this approximate treatment in the App "Drift_Diffusion" with an output shown in Figure 3.39. The electric field is chosen by "Slider" and results for mobility and diffusion r.m.s. are displayed for gaseous and liquid Argon. The drift distance and diffusion smearing are plotted for a gas with a drift

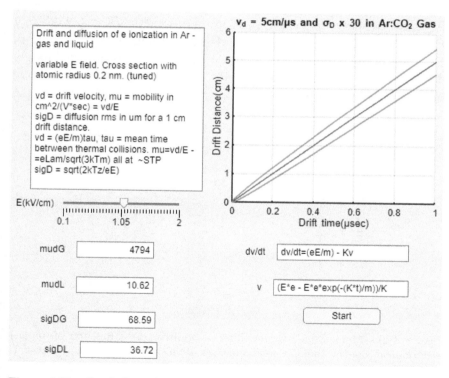

Figure 3.39: Symbolic solution for the velocity as a function of time. The mobility of Argon gas and liquid is shown in $(cm^2)/(Vs)$ units, while the r.m.s. of the diffusion distance for a 1-cm drift distance is shown in μm units. The plot is for a 1-μs drift time with a 5-cm/μs drift velocity and diffusion multiplied by 30 for visibility.

velocity of $5\,cm/\mu s$ as a function of drift time. The equation of motion is solved symbolically using "dsolve" which shows that the drift velocity attains a constant value in a time $\sim m/K$.

The mobility, μ, is proportional to the drift velocity but has the electric field factored out and is scaled to the density at STP. It is conventional to use cm, V and s units for mobility and drift velocity and that convention is followed here. For several gases commonly used in detectors the mobility is approximately in the range $1000\text{--}10000\,(cm^2/Vs)$. For an $ArCO_2$ (90:10) mixture the drift velocity is $\sim 5\,cm/\mu s$ for an applied field of $\sim 1\,kV/cm$. Data for LAr is shown in Figure 3.40. In this case the drift velocity depends strongly on additives to pure LAr. For $E < 0.3$ kV/cm the mobility is approximately field independent as expected from Eq. (3.34). The mobility for $E = 0.2$ kV/cm is approximately $400\,cm^2/(Vs)$ and the drift velocity is $\sim 0.8\,mm/\mu s$ which is slow with respect to gases because of the increased density of the material.

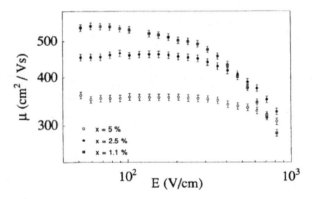

Figure 3.40:   Mobility in LAr for different additives as a function of applied electric field. Some commonly used additives are methane, ethane and ethylene.

At higher fields the approximation of ignoring the effect of the guiding field compared to the thermal velocity becomes invalid and the mobility typically decreases rather rapidly. The mobility behavior is complex and very dependent on the gas or liquid purity as well as the applied field which makes experimental determination of the drift

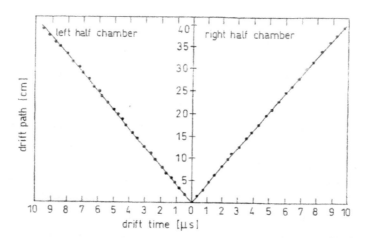

Figure 3.41: Data from a gas filled drift chamber. There is an ambiguity whether the ionization arrived from the left or the right side of the pickup. The ionization can drift to the anode wire either from the left or the right and have the same drift time. This ambiguity must be broken by providing multiple measurements with different pickup locations.

velocity a requirement. There are also additional complications if the drift occurs in the presence of a magnetic field which is often the case when magnetic spectrometers are employed. Data on drift times in a gas filled chamber appear in Figure 3.41. The drift velocity is about $4\,\mathrm{cm}/\mu\mathrm{s}$.

The drifting ionization will induce a signal on a pickup electrode which can then be used by electronic detectors. In this text only the initial signal formation and not the full electronic signal processing will be discussed. For example, the liquid argon detector has no gain in charge. Only the ionization charge, q_s, is measured which is called a "unity gain" detector here. For an idealized parallel plate geometry with voltage V_o applied to a gap of length d, the field E is constant, $E = V_o/d$, and the drift velocity is therefore constant. For a charge $q(t)$ moving across the gap there is a signal induced on the electrodes. The motion of the ionization induces a capacitive current on the electrodes. Since the field does work dU on the charge, energy conservation can be used, as in Eq. (3.36) where Q and C refer to the charge and capacity of the electrodes. The energy stored in the capacitor is U.

$$Q = CV, \ U = CV^2/2 = Q^2/2C$$

$$dU = QdQ/C = Fdz = q(t)E(z)[v_d(t)dt] \qquad (3.36)$$

$$I(t) = dQ/dt = q(t)E(z)v_d(t)/V_o$$

A charge, q_s, that appears at a point, z_o, in the gap will be referred to as point ionization. That might occur in an alpha particle decay, for example. The electron will move to the anode, at $z = 0$, at a constant drift velocity, $-v_d$, and will terminate at $z = 0$ in time z_o/v_d. The current induced by the electrons on the anode is $I(t) = q_s(v_d/d)$ and the total induced charge is $q_s(z_o/d)$. The ions from the point move to the cathode at a much lower drift velocity and therefore induce a smaller current on the cathode, of the same sign as the electron current.

A charge passing completely through the gap at right angles, with a length d, produces a line charge of ionization initially with total source charge $\lambda d = q_s$. The induced current at $t = 0$ is λv_d which falls to zero at $t = \tau$ when all the ionization is sunk into the electrode where τ here is the total drift time. Note that here and in what follows τ is defined to be the pulse formation time and not the mean time between collisions as it was used above in this section. The charge in the gap is $q(t) = \lambda d(1 - t/\tau)$, while the current on the electrode is $I(t)$ using Eq. (3.36). The charge refers only to the charge induced by the electrons. The ion current is ignored. For a 1-cm gap with a $1\,\mathrm{kV}$ voltage and an Ar–CH$_4$ mixture, the charge collection time is about $20\,\mu s$.

$$\tau = d/v_d$$

$$I(t) = \lambda v_d(1 - t/\tau)$$

$$Q(t) = \lambda v_d(t - t^2/2\tau) \qquad (3.37)$$

$$Q_{tot} = \int_0^\tau I(t)dt = (\lambda v_d \tau)/2 = \lambda d/2$$

For the electrodes one simply collects charge $Q(t)$ on "pads". The configuration of these pads is up to the specific application.

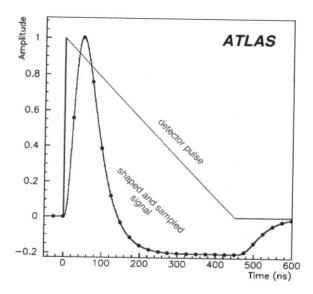

Figure 3.42: The electrode signal of a LAr detector showing the linear falloff in time as the line charge is sunk into the pickup electrode. For a more rapid response, about 50 ns, the signal is shaped by the front-end electronics.

The designer is free to choose the pad configuration to localize the incident charged particle. Charge sharing over electrodes can be used to further localize the incident particle. This feature will be explored in more detail when the segmentation of the cathode of a gas filled chamber is discussed in Section 3.16 but the principle is the same in all cases. The signal for a specific LAr detector is shown in Figure 3.42 in the case of full line ionization. The time constant is about 500 ns. This is a bit too long for the LHC application where it is used, so that the electrode signal is shaped by the electronics which sample the pulses.

3.16. Detector Gas Gain

Gas filled detectors are used so as not to interfere overly with the particle trajectory. For a unity gain device, many particles are needed in a beam to give a reliable signal above noise sources. That is the mode of operation of an ionization chamber. The collected charge is linearly proportional to the number of incident relativistic (mip) particles. However, detection of a single mip is difficult since the signal is small. In this case the ionization must be amplified. If the amplification is unconstrained the pickup itself is discharged, which is called "Geiger mode". Indeed, the SiPM discussed previously operate in that mode.

If the amplification, due to collisions with the electrons ionizing the gas, is limited and proportional to the incident ionization, the resulting device is called a Proportional Wire Chamber (PWC). The cascade of ionization is similar to the shower already mentioned in regard to electron interactions in high Z materials. It is not inoperative while being recharged as is a Geiger device making it capable of higher rate operation.

The PWC pulse formation with gain is due to anode signal amplification in the high field of a small diameter wire. The electrons sink onto the anode wire and induce only a small current. We consider only the simplest case of a single conductive tube of radius b with a central wire anode of radius a at positive potential V_o with charge per length λ. Using basic electrostatics the field in the tube and the voltage are, $E(r) = \lambda/(2\pi\varepsilon_o r)$, $V_o = (\lambda \ln(b/a)/(2\pi\varepsilon_o)$.

The primary ionization electrons drift to the central, positive, anode wire. They are accelerated and multiply by ionizing the $ArCO_2$ or other appropriate gas. For example, the mean first ionization potential in Ar gas is $26\,eV$. The multiplication happens near the wire, in the high fields. It is proportional to the initial ionization for low gains, of order 10^5, or it can totally discharge the wire in the Geiger mode, for gains above about 10^8. The situation here is one where the charge is proportional to the initial signal. The signal appears near the wire, effectively instantaneously as in the case of point ionization in the LAr discussion.

After a gain of the charge, the electrons sink to the wire while the ions move towards the tube wall, albeit more slowly. Since the electrons move only a short distance, their contribution to the signal current is small and is neglected. Compared to a LAr parallel plate detector there is a gas gain and the field is not constant, but the ionization can be considered to appear at a point near $r = a$. In Eq. (3.38) the mobility now refers to the positive ions not the electrons. The characteristic distance scale is the wire radius, a, and the characteristic time for the signal, τ, is approximately the time it takes the ions to drift, $v_d = \mu E(r)$, a distance $a/2$ while moving toward the cathode.

$$dr = \mu E(r)dt = \mu\lambda/(2\pi\varepsilon_o r)dt$$

$$\int_a^r r\,dr = \mu\lambda/(2\pi\varepsilon_o) \int_o^t dt \tag{3.38}$$

$$r(t) = a\sqrt{1 + t/\tau}, \tau = a/[2\mu E(a)]$$

The initial current pulse, $I(0)$, assuming $r = a$ for the ions at $t = 0$, is shown in Eq. (3.39). For example, for $b = 1$ cm, $a = 20\,\mu$m, $E(a) = 160$ kV/cm, $\lambda = 180$ pC/cm, so that with an ion mobility of $1.5\,\text{cm}^2/(\text{Vs})$ the pulse formation time τ is 8.3 ns. The ions drift for a long time, so typically one uses a current amplifier and shapes the pulse on a time scale approximately that of the pulse formation time. This shaping is similar to what was seen in Figure 3.42 for LAr. A schematic of the signal for a PWC is shown in Figure 3.43, where τ refers to the pulse shaping time and not the pulse formation time

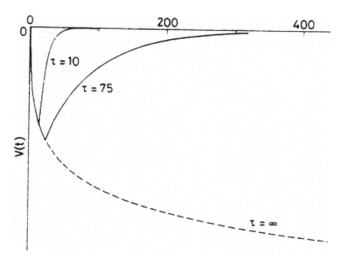

Figure 3.43: Schematic diagram of signal pulse in a PWC for different electronic shaping times. Typically, only the rapid initial pulse shape is used for signal processing.

which defines the initial time behavior of the pulse.

$$v(t) = (a/2\tau)/\sqrt{1 + t/\tau}$$
$$I(t) = -Q_{ion}E(r)v(t)/V_o \qquad (3.39)$$
$$= \sim -Q_{ion}[1/(t + \tau)]$$

The current signals, $I(t)$, of typical LAr and PWC are compared in Figure 3.44, created by the script "LA_PWC". A schematic LAr detector with $v_d = 2\,\text{mm}/\mu s$ and a 1-mm, gap with a 1 kV/cm field is assumed. For the PWC a gas gain of 10^4 was assumed with ionization deposited over 1 cm in Ar gas. The pulse formation time was 8 ns, $b = 1\,\text{cm}$ and $a = 20\,\mu\text{m}$. The PWC pulse is large for the first 50 ns while the LAr pulse drops linearly to zero at 500 ns. The differences in the electric field, constant for LAr and $1/r$ for the PWC, lead to the different temporal shapes. Typically, a LAr detector is used in shower detection where the number of particles making the signal is large. For PWC the gas gain more than offsets the factor ~ 1000 in density and single particles can be detected without changing their energy or velocity significantly. For that reason, PWC are often used as "tracking" detectors.

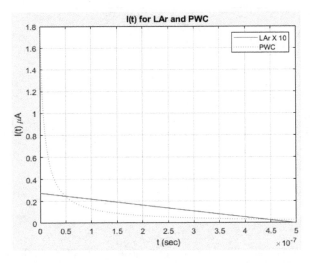

Figure 3.44:   Signals $I(t)$ for a schematic LAr and PWC detector. The gas gain of the PWC and the faster time structure of the PWC make the initial currents larger than the LAr.

There is also an induced charge on the cathode of a PWC as there is on both electrodes of a LAr detector. The cathode can be freely configured. One possibility is to use strips perpendicular to the wires in order to measure a second orthogonal coordinate in a single detector plane, the first being found using the drift distance. A parallel plate configuration is now assumed for mathematical simplicity. When several planes are deployed and the charges induced on the strips are recorded, as shown schematically in Figure 3.45, both the position and the angle of the particle trajectory can be determined. The cathode signal is spread over several strips and the centroid of the charge can be used to obtain a more accurate coordinate than would be obtained if only a binary hit/no hit record of the strips were made. The method of image charges is used to evaluate the transverse electric field at the strip which is proportional to the surface charge density at the cathode. The parallel field component vanishes as a boundary condition.

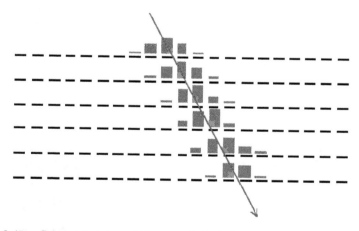

Figure 3.45:   Schematic view of the use of the induced charge on cathode strips for several PWC deployed at different z locations. The centroids of the induced strip charge, shown in green, are displayed as red crosses.

A more quantitative exploration of the use of cathode strips to find accurate particle coordinates is made in the App "CSC_Strip_Charge". The induced charge on the cathode is assumed to be in a central strip of 5 evaluated and is displayed symbolically. The charge is assumed to be the electrostatic solution. The strip width in x is 2w and is chosen by "Slider" when scaled to the wire distance above the cathode plane. The anode charge is at $z = d$, $x = x_q$, $y = 0$. The integral over the strip y is done using the Matlab symbolic utility "int" and is displayed in the App using an "EditField". The next integral over the strip x for a given x_q is also done symbolically and displayed. Both integrals are displayed as indefinite.

A specific output of the App appears in Figure 3.46. The user can vary w/d and see that a very small value means that the charge will spread beyond the 5 strips, while a large value means the sharing

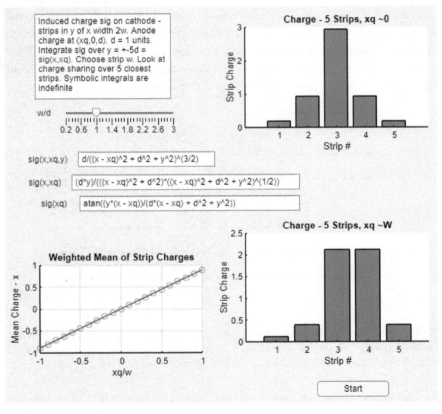

Figure 3.46: The cathode is in the (x, y) plane at $z = 0$ with strips in y of full width 10d where 2w is the full width of the strips in x. The charge is at the anode wire at $z = d$, $y = 0$ and $x = x_q$.

among the 5 strips is very minimal. A compromise is to have $w/d \sim 1$. The charge sharing for a signal at $x = 0$ and $x = w$ is plotted to display how the charge centroid is used to measure x_q accurately with sufficient strip sharing. The weighted mean of the strip charges is shown over the full range of the third strip centered on $x = 0$. The user is encouraged to look for deviations from the true value of the charge location as w/d is varied. This cathode configuration is a simple Cartesian one. Others can be created and studied using the symbolic and numeric tools supplied by Matlab. For his work on PWC and other detectors Georges Charpak was awarded the Nobel Prize in Physics in 1992.

3.17. Solid State Detectors

In the last few years, silicon detectors have become the detector of choice in many applications as the improvements in cost and availability have been enormous, profiting from the explosive growth in the semiconductor industry. Isolated atoms have sharply defined quantized energy levels as previously noted. However, in a solid the electrons of different atoms interact, and those sharp levels become "bands" of allowed and forbidden energies. In metals there are allowed states with nearby vacant states so that the electrons are free to move. In insulators the highest energy band is full, and the next allowed band is empty and separated by a large energy gap which inhibits thermal excitation of electrons. Semiconductors are elements intermediate between metals and insulators.

Silicon is a semiconductor with an energy band gap between the top of the filled valence band (VB) and the bottom of the empty conduction band (CB) of 1.12 eV. Since that is $\sim$44 times the value of kT at 300°K $\sim$ 0.025 eV, there are very few thermally excited electrons in the CB. The intrinsic number of thermal charge carriers is only about $1.5 \times 10^{16}/m^3$ which is about 10^{-12} of the density of atoms of silicon, $\sim$5 $\times$ 10^{28} atoms/m^3. The intrinsic resistivity is measured to be, $\rho_i \sim 2.3\,k\Omega m$. It takes 3.62 eV, a measured energy greater than the band gap, to make a hole-electron pair in silicon due to the existence of other competing effects.

Silicon $300\,\mu$m thick has 116 keV deposited by a mip which then creates 32,000 electrons in the CB or a signal of 5.1 fC. In the CB the electrons have a measured mobility of $1400\,\text{cm}^2/(\text{Vs})$, while the holes or vacancies remaining in the VB are about three times slower, $450\,\text{cm}^2/(\text{Vs})$. The permittivity, ε, of silicon is $\sim$1 pF/cm which is a factor 11.4 times the vacuum permittivity of 8.85 pF/m.

A diode is formed by doping the pure silicon with small densities of impurities that have quantum states between the CB and the VB in the forbidden band gap. Impurity states near the bottom of the CB can donate, p type, thermally excited electrons easily, while those with states near the top of the VB, n type, can accept electrons from the VB easily leaving holes. The two types are then combined together. There are fixed ions, and mobile charge carriers due to these impurity states. For p type the ions are positively charged, while for n type they are negatively charged. The ions create an electric field with direction going from n type to p type.

If a voltage is applied in "reverse bias" the mobile carriers are swept up by the electrodes, leaving just the ions. That is a positive voltage on the n side or a negative voltage on the p side. Assume the p side, called the "bulk" in this configuration, extends a length along the z axis of d and is lightly doped, while the n side is much narrower and much more heavily doped, typically by a factor $\sim 10^3$, in order to preserve charge neutrality. The width of the p type is then typically 10^3 times larger than that of the n type or vice versa for the other type of diode. If all the mobile carriers are removed by the reversed bias voltage, the diode is then "fully depleted" and has a very small thermal current of the minority carriers. That fact makes the diode a good particle detector, sensitive to the very small 5.1 fC signal of electron/hole pairs. A schematic of a diode is displayed in Figure 3.47.

Passage of a charged particle through the diode will create electrons and holes. The holes, $q > 0$, will be swept to the negative electrode, while the electrons, $q < 0$, are swept to the positive electrode, the n type. For the PMT and the PWC, which are collecting electrons, the anode is positive with respect to the cathode. The holes drift to the p side while the electrons drift to the n side.

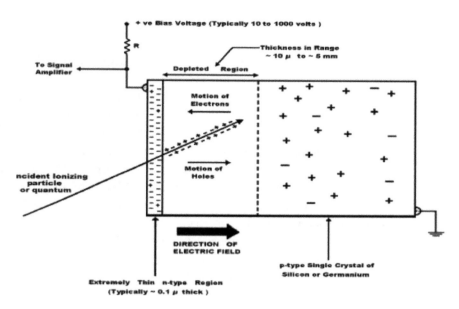

Figure 3.47: Schematic diagram of a diode composed of a thin, highly doped n type coupled to a thick lightly doped p type the "bulk", with a positive voltage applied to the n type which causes a region depleted of mobile charge carriers.

The voltage arrangement can be reversed. In what follows a diode where the "bulk" is n type and not p type as in Figure 3.47 will be explored. The choice of the bulk is made by the needs of a particular experiment and both types will be mentioned here.

The electric field in the diode is linear in z which follows from the assumed uniform density of ions, $\partial E/\partial z = \rho/\varepsilon, E(z) = (\rho/\varepsilon)(d - z), E(0) = \rho d/\varepsilon$. The charge density, ρ, is qe times the number density, n. An external voltage V_o is applied across the diode. $E = -\partial V/\partial z, V(z) = -(\rho/\varepsilon)(dz - z^2/2), V_D = -\rho d^2/2\varepsilon$. The other diode type has a "bulk" n side and a thin p side. In that case electrons still move the n side, the "bulk", electrode and holes to the p, thin, side. Approximating the p side as very thin and ignoring the diode surface voltage of about 0.75 V compared to the applied voltage, the depletion voltage is taken to be the voltage when the n region is fully depleted, when there is an electric field throughout the entire detector. The field just at depletion is $E(z) = 2V_D(d - z)/d^2$. In some designs the detectors may be operated at over depletion, with a non-zero electric field existing at $z = d$. This is done to speed up the time response of the detector, but this situation will not be treated here.

The solution for the charge motion as a function of time is exponential, reflecting the fact that the electric field is linear in z for this detector type, in distinction to the LAr and PWC types discussed previously. In what follows a bulk n type is assumed. In the specific case of point ionization at z_o, the results in Eq. (3.40) for position and velocity of the electrons and holes are obtained. Since the electron and hole mobilities are now similar, the induced currents of both on the anode are followed. Only the drift in the "bulk" material is treated as the other type is considered to be vanishingly thin. The electrons move from z_o to d, the holes to $z = 0$. Since the field vanishes at $z = d$, both the velocities vanish there.

$$E(z) = -[2(d - z)/d^2]V_D$$

$$\tau = d^2/(2\mu V_D) = d/(\mu|E(0)|)$$

$$dz/dt = \mu E \tag{3.40}$$

$$[(d - z_e)/(d - z_o)] = e^{-t/\tau_e}, v_e = (e^{-t/\tau_e}/\tau_e)(d - z_o)$$

$$[(d - z_h)/(d - z_o)] = e^{t/\tau_h}, v_h = -(e^{t/\tau_h}/\tau_h)(d - z_o)$$

For a concrete example consider a 300-μm thick detector, lightly doped with resistivity $\rho = 0.06$ kΩm which is 38 times smaller than the intrinsic value. The depletion voltage is 54 V and the capacity per area is $\sim$35 pF/cm^2 or a source capacity of 8 pF for a 5×5 mm^2 detector pixel. The source capacity will be of importance in a later discussion of noise sources, since the signal is quite small in this unity gain device. The electrons drift to $z = d$, while the holes drift to $z = 0$ albeit about three times slower. At the electrodes, the induced current pulses follow from the same considerations which have been used for the unity gain LAr detectors treated previously except that the field is now position dependent. The characteristic time for charge collection is, Eq. (3.40), $\tau = d/[\mu E(0)] = d^2/(2V_D\mu)$. For point ionization at z_o the electron and hole currents are shown in

Eq. (3.41) where the exponential factor reflects z the dependence of the E field. A lower depletion voltage means a longer time constant, $\tau \sim 1/V_D$.

$$I_e(z,t) = (q_s/\tau_e)[e^{-t/\tau_e}(d - z_o)/d]$$
$$I_h(z,t) = (q_s/\tau_h)[e^{t/\tau_h}(d - z_o)/d]$$

(3.41)

The appearance of a mip particle at z_o at $t = 0$, creates ionization. The hole and electron currents vanish at $z_o = d$ because there is no electric field. The case for a through-going line of total ionization is treated by integrating over all points z_o from (0,d) assuming a constant line density, λ, with total source charge $q_s = \lambda d$. The current pulse for electrons and holes with line ionization appears in Eq. (3.42). The current at the anode reflects both the hole motion to the cathode and the electron motion to the anode.

$$I(t) = I_e(t) + I_h(t) = \lambda \int_0^d I(t, z_o)dz_o$$
$$I(t) = (\lambda d/2)[e^{-t/\tau_e}/\tau_e + e^{-t/\tau_h}/\tau_h]$$

(3.42)

The collection time for the example used above is $\tau_e \sim d/[\mu E(0)]$, about 6 ns for electrons (18 ns for holes) and the maximum electron drift velocity is $\sim 50\,\mu m/ns$. The peak electron current is $q_s/\tau_e \sim 800\,nA$, while the peak field is $\sim 3.6\,kV/cm$. Shorter collection times can be achieved by operating the devices above the depletion voltage. A diode with a 300-μm thickness is shown using the script "Si_Diode_App_2" with output shown in Figure 3.48. The basic choice is the doped resistivity which is chosen by "Slider". The initial App setting corresponds to the example given in the text above. Given d and ρ, the depletion voltage, the electron time constant, the drift velocity, the maximum field and the maximum current are found and displayed using "EditFields". The solutions for the time dependence of z for e and h Eq.(3.42) are shown symbolically. The initial z value, z_o, for point ionization is also chosen by "Slider". The subsequent time development of z for both holes and electrons is plotted as a function of time in units of τ_e. The current for line ionization for electrons and holes is also plotted. The choice of resistivity defines

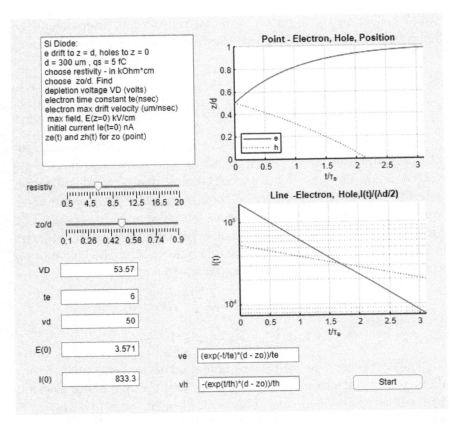

Figure 3.48: Output of the script "Si_Diode_App_2". The user chooses the resistivity of the lightly doped p type and assumes the n type is very highly doped and therefore thin.

the electrical characteristics of the diode, while the choice of initial z value for the point ionization gives a graphic way to see how charge is collected throughout the thickness of the diode.

Data on a Si diode current pulse is shown in Figure 3.49. The time scale for the rise time and the full width and half maximum, FWHM, illustrate that these devices have ns time resolutions. The more complex semiconductor deployment of a typical "microstrip" detector is shown in Figure 3.50. The bulk in n type in this example. Individual strips can be read out to provide quite precise localization of a particle in alalogy to the operation of the cathode strips in a

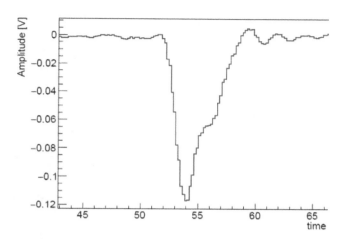

Figure 3.49: Electron current pulse from a Si diode. The rise time is sharp, $\sim 2\,\text{ns}$, and the width of the pulse is comparable to that for typical diode electron time constants, although shaping has been performed. The horizontal time scale is nanosecond.

PWC. Pitch distances between strips of the order 90 μm are common and the effective resolution of the devices is greater than that of the cathode strips of PWC because the feature size of the strips is much smaller. Further advances in the microelectronics industry can also make more complex devices available to the research community.

Why is noise an important consideration for most detectors? If noise is thought of as a fluctuating small signal then a real signal measurement of time and amplitude has uncertainties. In turn, that means, for example that TOF and dU/dz measurements of particle passage time and energy deposit have noise errors. If the signal is small, noise is of particular concern. For the silicon diodes the reverse current of minority carriers is a noise source which can be reduced by lowering the temperature and such detectors are often run at low temperatures. Specifically, radiation damage to the detectors creates lattice defects which are realized as impurity states between the VB and the CB. These states increase the reverse current and that current constitutes another noise contribution to the signal current.

Principles of operation

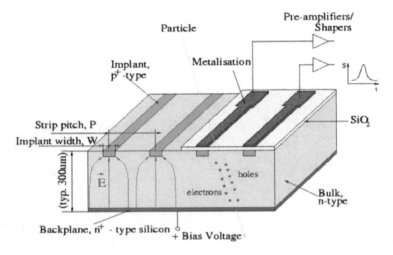

Figure 3.50:   Setup for typical microstrip detector; thin p type, $300\,\mu$m of n type "bulk", strips of p type. There is an ion implant, $n+$, for the backplane electrode.

The signal is represented as a current source I_s. That ideal current source is modified by a source resistance, R_s, and a shunt capacity C_s. Noise has a flat frequency spectrum, or a delta function time spectrum as will be further explored in Chapter 4. To limit the noise some frequency filtering is therefore needed. Thermal noise in the source resistor per Hz of bandwidth is $dI_T^2/d\omega = 2kT/\pi R_s$. It can be reduced by lowering T or increasing R_s. The shot noise due to fluctuations in the discrete charge carriers, electrons of charge e, making up the base current of an assumed front-end transistor is $dI_s^2/d\omega = eI_b/\pi$. It can be reduced by lowering the standing current I_b but that may slow the transistor response. There is also front-end transistor noise characterized by the trans-conductance, g_m. These noise sources are represented in Eq. (3.43) as a squared noise voltage. The needed frequency limiting is provided here by a simple filter model, $f(\omega)$, which has gain G and a time constant τ which limits the voltage at both high and low frequencies. Integrating $dV^2/d\omega$ over ω, the mean square noise voltage, $\langle V^2 \rangle$, is finite due to the cutoff provided by $f(\omega)$. The numerical values in $\langle V^2 \rangle$ depend on the filter details, which can be tuned, but the basic functional dependence is fixed.

$$dV^2 = [kT/2R_s(\omega C_s)^2 + eI_b/(\omega C_s)^2 + 2kT/g_m]d\omega/\pi$$

$$f(\omega) = G[\omega\tau/(1+(\omega\tau)^2] \qquad (3.43)$$

$$\langle V^2 \rangle = G^2[(kT/2R_s + eI_b)\tau/C_s^2 + 2kT/(g_m\tau)]$$

The noise sources are referred to the input so that the signal to noise performance can be directly evaluated. This procedure is shown schematically in Figure 3.51, where the thermal noise in the source resistor is represented as voltage source at the input with the source resistance and capacitance. The other noise sources are also referred to the input.

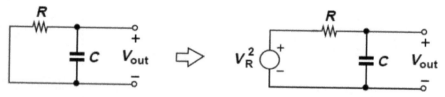

Figure 3.51:   Schematic representation of how the thermal noise in the source resistance is represented as a voltage source at the detector input.

The signal at the peak frequency of $\omega = 1/\tau$ set by the filter is $V_s \sim (q_s G/(C_s e_{\ln}))$ where $e_{\ln}$ is the natural exponential constant. The equivalent noise charge due to the three noise sources, referred to the input signal, is shown in Eq. (3.44). It is customary to define parallel and series noise voltages, V_P and V_S which give equivalent parallel and series noise charges (ENCP and ENCS) which scale as $\sqrt{\tau}$ and $1/\sqrt{\tau}$, respectively. The factor e in the parallel voltage is the electron charge, not to be confused with $e_{\ln}$. Note that the gain has been factored out.

$$ENCP = (C_s e_{\ln}/G)\sqrt{\langle V_P^2 \rangle} = (e_{\ln}/2)\sqrt{\tau\,(kT/2R_s + eI_b)}$$

$$(3.44)$$

$$ENCS = (C_s e_{\ln}/G)\sqrt{\langle V_S^2 \rangle} = (e_{\ln}/2)C_s\sqrt{2kT/(g_m\tau)}$$

To reduce noise one wishes low T, large R_s (ideal current source), low base current (conflicts with high speed operation), small source capacity (limits design of strips) and a compromise for the filter τ which optimizes the series and parallel noise. The optimization of the noise is specific to each individual application. Numerical values for the thermal resistor noise and the base shot noise are given in Eq. (3.45) for reference. As a numerical example, the previous silicon detector, 300 μm thick, has a 5.1 fC source charge of 32,000 electrons. Using a transconductance $= 1/g_m$ characteristic of a forward biased diode, $g_m = 1/25\Omega^{-1}$, a source capacity of 10 pF due to leads and not the 35 pF/cm^2 of the diode itself if the strip area is small, a 10 kΩ source resistance, a 10 μA base current and a 10-ns filter shaping time (to match the signal collection time), at STP the ENCS is 1340 e, the ENCP is 385 e and the signal to total noise, S/N, is $\sim$23, where N is ENCS and ENCP added in quadrature.

$$\sqrt{2kT/R} = 0.091nA\sqrt{Hz/\Omega}$$
$$\sqrt{eI} = 0.4nA\sqrt{I(A)Hz}$$
(3.45)

Systems of solid state detectors are deployed, for example as concentric cylinders surrounding an interaction point in a solenoidal field. The momentum and charge are inferred by the curvature in the azimuthal direction. The maximum detectable momentum arises when the measurement error of the detectors makes the curvature un-measurable, since $dp/p \sim p$. The minimum momentum limit is set by multiple scattering in the detector elements since dp/p in that case is a constant.

Very precise tracking with solid state detectors is needed to detect the decay vertices of particles containing b quarks. The masses of such particles are about 5 GeV and they have lifetimes, at rest, typically $c\tau \sim 450\,\mu$m. Time dilation in SR increases the observed lifetime by a factor γ of the moving b quark, making effective decay distances longer by that factor. Therefore, a solid state "pixel" with linear dimensions $\sim$150 μm is needed to distinguish the primary production vertex of the particle from its decay vertex, thus identifying that a heavy flavor particle has decayed. The observation of an impact parameter resolved from the primary vertex defines a decay candidate. A schematic of the primary and secondary vertices

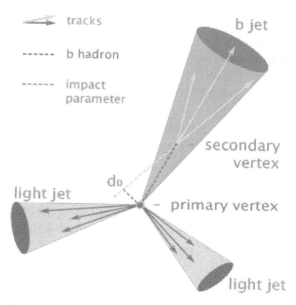

Figure 3.52:   Schematic diagram of the topology of a heavy flavor b decay. The primary and secondary decay vertices are shown and the definition of the impact parameter, d_0, of a decay track is indicated.

is shown in Figure 3.52. The resolution of a silicon detector of pitch d in simple binary, yes/no, operation is $d/\sqrt{12}$. More accurate position measurements are possible if the ionization signal itself is recorded rather like the situation discussed with PWC cathode strips where the charge centroid was used for a more accurate position determination.

Recently solid state detectors with gain have been developed, analogous to the gain attained in PWC. The electric field is enhanced within the the detector. With a large enough field there is charge multiplication by impact ionization, but not Geiger discharge, as in a PWC. These detectors offer very good timing measurements because of the enhanced S/N performance. For example a 30 ps time resolution corresponds to a UR particle location to 3 mm. That level of timing opens up the possibility of "4D" tracking, or tracking in (x, y, z, t). Such tracking opens a new dimension that aids in improving the operation of tracking systems in high rate, high luminosity environments, such as the upgraded LHC accelerator will soon be providing.

Another recent development in the use of silicon detectors has been in the realm of electromagnetic calorimetry. In the past liquid argon, crystals, or scintillator have all been used as the active detector media. However, liquid argon is fairly slow and crystals and scintillator do not offer the very fine granularity needed to explore the details of the development of the shower.

3.18. Coaxial Cables

All these instruments need to be connected together, which is a mundane but very important issue in deploying complete systems. A common solution is to use coaxial cables. These cables have a large bandwidth and are reasonably immune to external electromagnetic noise. A more recent use of optical fiber cables has occurred because they are even more immune and provide electrical isolation between transmission and reception. However, they will not be discussed in detail here, only cursorily at the end of this Chapter.

A coaxial cable is basically a cylindrical waveguide loaded with a dielectric. In a conductor an electric field which oscillates can penetrate a depth called the skin depth, δ. The value of the skin depth depends on the conductivity, $\sigma = 1/\rho$ where ρ is the resistivity, the permeability, μ, of the material and on the frequency, ω, of the field. In an ideal conductor the conductivity is purely imaginary which suppresses the field exponentially. The units of permeability μ are $\Omega s/m$, ρ are Ωm and ω are $1/s$.

$$E \sim e^{-z/\delta}, \delta = \sqrt{2\rho/\mu\omega} \qquad (3.46)$$

For example, in copper, $\rho = 1.7 \times 10^{-8}\,\Omega m$; taking $\mu = \mu_o$, the skin depth is 6.6 cm divided by $\sqrt{f}$ in Hz. At 60 Hz (house power) the depth is 8.5 mm. That allows one to simply twist wires together at home to make a sufficiently good contact. At 100 MHz the skin depth is 6.5 μm or 0.65 μm for $f = 10\,GHz$ and the fields do not penetrate any reasonable thickness of copper. Clearly, efficient power transmission is easier at high frequencies.

Turning to a cylindrical coaxial cable with inner radius a, and outer radius b, the static field was already noted in Eq. (3.38) in the PWC discussion. The cable is filled with material with permittivity ε, where the vacuum permittivity is $\varepsilon_o = 8.8\,\mathrm{pF/m}$. The permeability is assumed to be $\mu_0 = 4\pi \times 10^{-7}\Omega\mathrm{s/m}$. The coaxial impedance depends on the cylindrical geometry, a and b, and on the dielectric between the inner and outer radii. The wave velocity is c/n and the impedance, capacitance (C_L), inductance (L_L) and resistance (R_L) per unit length are all shown in Eq. (3.47).

$$v = c/\sqrt{\mu\varepsilon} = c/n$$
$$Z_o = \sqrt{L/C}$$
$$C_L = 2\pi\varepsilon/\ln(b/a) \qquad (3.47)$$
$$L_L = (\mu/2\pi)\ln(b/a)$$
$$R_L = (\rho/2\pi\delta a)$$

The impedance of the vacuum is shown in Eq. (3.48) and is $377\,\Omega$. For a coaxial cable the numerical value is shown in Eq. (3.48), again assuming $\mu = \mu_o$. A standard cable is called RG58 which has $50\,\Omega$ impedance. The bandwidth of the cable is finite due to resistance, Eq. (3.47), caused by finite skin depth δ. For RG58 the rise time of a step function pulse is $\sim 0.2\,\mathrm{ns}$, or a bandwidth $\sim 5\,\mathrm{GHz}$ for $10\,\mathrm{m}$ of cable. In most applications this bandwidth is sufficient for short cable lengths. The R_L value is $(1/2\mathrm{a})\sqrt{\mu\rho f/\pi}$ which shows that compact cables have more limited bandwidth than thicker cables.

$$Z_{vac} = \sqrt{\mu_o/\varepsilon_o} = 377\Omega$$
$$Z_o = (1/2\pi)\sqrt{\mu/\varepsilon}\ln(b/a) = 60\Omega\ln(b/a)/\sqrt{\varepsilon/\varepsilon_o}] \qquad (3.48)$$

Numerically, for a cable with copper conductor the skin depth is $\delta = 16.5\,\mathrm{cm}/\sqrt{(\omega(\mathrm{Hz}))}$, for an inner radius $a = 0.1\,\mathrm{cm}$ and $R_L \sim 10^{-7}\Omega/(\mathrm{cm}\sqrt{(\omega(\mathrm{Hz}))}$. For a 10-m cable at $\omega = 1\,\mathrm{GHz}$ the resistance is $3\,\Omega$ and the capacitance is $300\,\mathrm{pF}$. A step function has a rise time $\tau \sim \mathrm{RC} = 0.9\,\mathrm{ns}$. Cable loss as a function of frequency appears in Figure 3.53. It is clear that cables are not a good choice for

Loss v Frequency Comparison for Fiber & Coaxial Cable

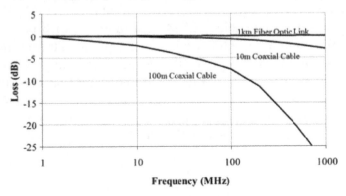

Figure 3.53:   Signal loss for coaxial cable as a function of frequency at 10 m and 100 m length compared to 1 km of fiber optic cable.

applications where GHz bandwidth is desired or where long cables are required.

The advantages of fiber optics to transmit over long distances with high bandwidth and immunity from electromagnetic noise are clear. Such fibers are also used to transmit signals in sensitive detectors since they present less mass to particles which transverse them and therefore introduce less scattering and energy loss. In fact, the 2009 Nobel Prize in Physics was awarded for research in fiber optical communication. The modulated signals are often created using LEDs which emit light in the IR range. The optical fibers can be very compact. The received light is converted back to an electrical signal using a photodiode, which uses the photoelectric effect. At present a figure of merit for a typical high-quality fiber is ∼3.5 GHzkm at 850 nm. It is clear why fiber optics has revolutionized the way telecommunication signals are sent over long distances. On the other hand there is a niche for coaxial cables because of their ease of use for short runs that require modest bandwidth performance.

3.19. Particle Tracking — x, α, p

Many of the instruments mentioned above are used in combination in a beam line or experiment. A good example is the use of position detectors in a magnetic field. For example, consider silicon detectors deployed in several planes at fixed z locations and measuring coordinates x when immersed in a dipole field, uniform in the y direction. A charged particle traversing these planes allows a measurement of q, p, x and dx/ds. Measurement error, as will be seen, limits the accuracy of the p measurement at high p values, while multiple scattering limits the measurements at low p values. If other systems such as TOF or Cerenkov counters are deployed the particle mass can also be determined.

In Eq. (2.10) the bend angle in a dipole of length L and field B was defined. In GeV, m, T units for p, L, and B the charge e is $= 0.3$ (Eq. 2.10). A transverse momentum impulse can be ascribed to the field and to the multiple scattering in material, Eq. (3.14). These results are repeated in Eq. (3.49). A picture of the tracking in a solenoidal field is shown in Figure 3.54. The positions of particles in a series of cylindrical detectors, inner of silicon, outer of gas filled PWC are shown as dots. The fitted tracks are shown as circular

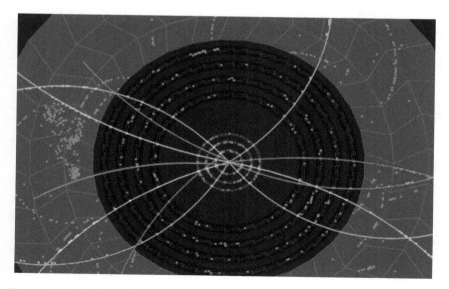

Figure 3.54: An interaction at the ATLAS experiment at the LHC. The tracking uses silicon and gas detectors and fits to helices emanating from the point-like interaction region.

trajectories.

$$\phi_B = L/\rho = (\Delta p_T)_B/p, (\Delta p_T)_B = qeLB$$

$$\rho = p/(qeB) \tag{3.49}$$

$$(\Delta p_T)_{ms} = (\varepsilon_s/\beta c)\sqrt{dz/X_o}$$

For a set of tracking detectors deployed at fixed z locations and measuring the x position of a charged particle moving in a dipole field $B_y = B_o$, the bend angle is measured with precision limited by the spatial resolution of the detectors. Assume there is an r.m.s. error of σ for all detectors. If the length of the field is $\sim L$, then the angular error is $d\phi_B$ is $\sim\sigma/L$. A rough estimate of the error in the measurement of p is shown in Eq. (3.50). A fixed angular error leads to a fractional momentum error which increases with p which limits the accuracy of high momentum measurements. At the low end of p, multiple scattering sets a limit to the fractional accuracy, dp/p, which occurs when $(\Delta p_T)_B \sim (\Delta p_T)_{ms}$. That situation occurs for $\beta \sim (\varepsilon_s\sqrt{l/X_o})/(ceLB)$ where l represents the material of the detectors. For that reason, the resolution of the system as a function of momentum is often parameterized as having a term proportional to p, due to measurement error, and a constant term, representing the effects of detector multiple scattering.

$$dp/p \sim p[(\sigma/L)/(\Delta p_T)_B]$$

$$dp/p = \sqrt{ap^2 + b} \tag{3.50}$$

For the specific example of five planes of silicon detectors spaced in z by 5 cm with resolution σ, in x in μm chosen by "Slider" immersed in a field, B_y, of 2 T was created by a short Monte Carlo App called "Circle_Fit_B_p". Track momentum, p, in GeV was also chosen by "Slider". The hits were generated with a track having initial x and dx/dz of zero. The fit allows for three free parameters, initial x and dx/dz and p. The Matlab utility "fminsearch" was used to minimize the r.m.s. of the Gaussian smeared data points from the points found by the best fit. No multiple scattering was applied at the momentum ranges of interest here. Specific output of the App is shown in Figure 3.55. The plot shows the initial points and the

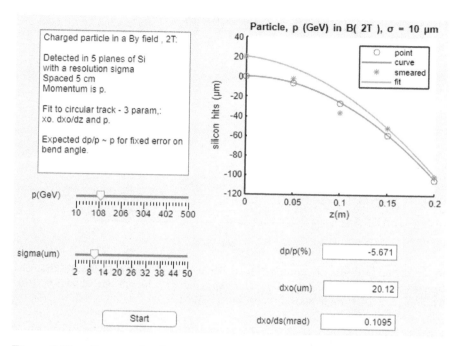

Figure 3.55: Output of a fit to a simple Monte Carlo model of a set of silicon detectors immersed in a magnetic field. The best fit and the actual trajectory can be compared visually for individual tracks.

curve they followed, along with the points smeared by the Gaussian resolution and the best fit to those points. The numerical values of the fit parameters, dp/p in percent, the initial x in μm and the initial dx/dz in mrad are displayed using "EditFields". The user should run a few examples for each choice of p and σ values to get a feeling for what the fit is doing.

The results of running 500 trials at a few different momenta and recording the r.m.s. is instructive. The results are plotted in Figure 3.56. The dependence of the three parameter fit shows that dp/p scales as p to the power 0.28 rather than 1. The reason for that behavior is that the fit assumed equal errors at each z location, and that the initial position and angle are free parameters, while the momentum is merely a third parameter. Given a particular set of Gaussian errors for a specific track, it becomes more advantageous

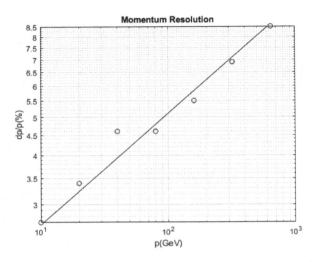

Figure 3.56:   Results of running 500 tracks at several momenta for the momentum resolution as a function of p. The exponent of p in the line shown is 0.28, while a value of 1 might naively have been expected.

for the fit to choose initial x and dx/dz values which are non-zero. If there were a fixed interaction point that was very well measured that could be assigned to the track, a different power of p would be observed. Indeed, $dp/p \sim p$ is only a first approximation. In the script the transverse momentum impulse, $(\Delta p_T)_B$, is 125 MeV and the resolution in angle is 0.05 mrad assuming the full length of 20 cm for the five planes. The approximate dp/p resolution is then $\sim 4\%$ at $p = 100\,\mathrm{GeV}$, which is not terribly far from the actual fit result of $\sim 5\%$ seen in Figure 3.56.

3.20. General Purpose Experiments

Many experiments deploy a whole suite of detectors in order to get more complete information on the reactions in which they are interested. They are called "general-purpose" experiments. Figure 3.54 shows an event from an LHC experiment, ATLAS, showing the tracking system. The reaction produces many charged and neutral particles which emanate from a very small interaction region. The charged particles curve in the axial magnetic field and their ionization is sampled in a set of silicon detectors followed by a set of gas filled detectors, such as PWC or drift chambers.

At larger radii, an electromagnetic calorimeter, ECAL, then is often installed to identify the showers due to electrons or photons and to measure their energies and angles. Outside of the ECAL a hadron calorimeter, HCAL, measures the energies and angles of the produced particles which are "strongly interacting". Finally, surrounding the whole experiment there may be "muon detectors" which measure the muon energy and angle, all the other particles having been absorbed in the calorimeters.

A schematic of the CMS experiment at the LHC appears in Figure 3.57. The basic systems are the tracking, the ECAL, the HCAL, the solenoid coil, and the muon tracking. The typical signals for both charged and neutral particles, which are distinctive, that are produced are indicated.

This concludes the Section which is specifically dedicated to detector and beam line instrumentation at an introductory level. The exposition has been a bit telegraphic of necessity — broad but not deep. Since whole texts have been written on any one of the 20 sub-topics, the reader is encouraged to first work with the scripts and Apps. If questions remain, use of the resources available on the

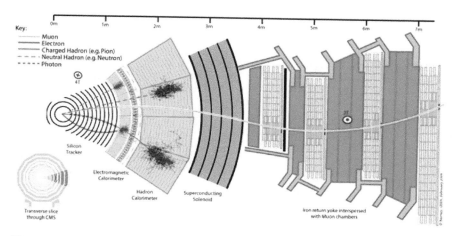

Figure 3.57:   Schematic view of an azimuthal section of the CMS experiment at the LHC.

internet to explore a specific subject in more depth can be most helpful. This text now turns to the topics relevant to accelerator instrumentation, where the issues are somewhat different, although related, to what has already been explored. The physics is unitary, but the specifics will be different.

Chapter 4

Accelerator Instrumentation

"It might seem paradoxical that the biggest scientific instruments of all are needed in order to probe the very smallest things in nature. The micro-world is inherently 'fuzzy' — the sharper the detail we wish to study, the higher the energy that is required and the bigger the accelerator that is needed."

— **Martin Rees**

"A linear accelerator has the advantage that no magnet is required and that its cost should not rise much more steeply than with the energy of the particles required."

— **Ernest Walton**

"It is impossible to travel faster than the speed of light, and certainly not desirable, as one's hat keeps blowing off."

— **Woody Allen**

4.1. Summary of Accelerator Parameters — α, β, γ, ε

The distinction between the topics in Chapter 3 and this chapter is that the instrumentation is typically outside the vacuum in the former and inside in the latter. There are many excellent reference books on accelerator physics and a few references are supplied in the section on references. In this text only a very brief introductory discussion is made. The emphasis later will be on instrumentation, so that the introduction made here is very schematic. In earlier chapters, matrix multiplication was used in MATLAB Apps to illustrate particle motion in uniform electric and magnetic fields. The dipole was discussed and a matrix representation for particle motion through a quadrupole was displayed. Following that a quadrupole doublet was shown to focus in both transverse directions and several focus conditions for both thick lens and thin lens quadrupole doublets were explored. That treatment was sufficient for use with beam lines, where a given particle passes a magnetic element only once. In discussing accelerators a different formalism is needed.

Accelerators will be assumed to be simple periodic structures. The existence of problems of injection and extraction of beams, special insertions and collision regions will not be addressed although they are in fact of primary importance in the operation of accelerators. For a periodic system with a longitudinal reference trajectory defined by a length s, a transverse coordinate $x(s)$ is assumed to have a quasi-harmonic behavior defined by a scale factor A, an amplitude function, $\beta(s)$, an oscillation phase $\psi(s)$ with an initial phase at s $= 0$, ψ_o. This ansatz, when substituted into Hill's equation for a periodic system subject to focusing, $k(s)$, which changes with s, $d^2x/ds^2 + k(s)x = 0$, $k = (1/B\rho)(\partial B(s)/\partial r)$, is indeed found to be a solution. If k were constant the solution would be simple harmonic motion. The dimensions of k are inverse length squared with a quadrupole magnet an example, Eq. (2.16).

$$x(s) = A\sqrt{\beta(s)}\cos(\psi(s) + \psi_o) \qquad (4.1)$$

It can be shown that a particle with transverse coordinate x and divergence $dx/ds = x'$, in traversing a periodic system of circumference C changes in a fashion describable by three parameters, α, β, and a phase μ. A fourth parameter, γ, is not independent but is conventionally used. The phase, μ, increases monotonically in traversing the circumference. It need not be periodic, a multiple of 2π, however. In fact, such a situation would be unstable. The tune is defined to be $Q = \mu/2\pi$. The lattice parameters α, β, and γ are periodic with a periodicity that corresponds to the periodicity of the focusing structure.

From Eq. (4.2), the α and μ parameters are dimensionless. The dimension of β is length and that of γ is inverse length. The matrix M is the transfer matrix connecting the transverse coordinates between s_o and $s_o + C$. It has determinant of 1, $|M| = 1$, and has eigenvalues and eigenvectors which can be determined symbolically using the Matlab utility "eig". If the matrix is known, $\beta_o = M(1,2)/\sin\mu$,

$\alpha_o = M(1,1) - 1/\tan\mu$ and $\cos\mu = \text{Tr}(M)/2$ where Tr is the matrix trace equal to the sum of the eigenvalues. It is perhaps confusing that in previous chapters the symbols α, β, γ, and ε were used for the fine structure constant, v/c, ε/mc^2, and energy. In addition, C was used for capacitance while here it is the circumference of a circular machine. The context of the specific text should make the confusion minimal. The symbolic conventions adopted in both cases are conventional.

$$\begin{pmatrix} x \\ x' \end{pmatrix}_{s_o+C} = M(s_o + C | s_o) \begin{pmatrix} x_o \\ x'_o \end{pmatrix}_{s_o}$$

$$= \begin{pmatrix} \cos\mu + \alpha_o \sin\mu & \beta_o \sin\mu \\ -\gamma_o \sin\mu & \cos\mu - \alpha_o \sin\mu \end{pmatrix} \begin{pmatrix} x_o \\ x'_o \end{pmatrix}_{s_o} \quad (4.2)$$

$$\mu = \psi(s) - \psi_o = \int_{s_o}^{s_o+C} ds/\beta(s)$$

The accelerator is assumed to be a conservative system and therefore, invoking the Liouville theorem, has a conserved "emittance", ε, which defines the phase space boundary in (x, x') space within which particles may orbit stably, or within some limiting aperture. It is assumed that a distinct boundary in (y, y') phase space also exists, and that there is no coupling between x and y motion in the simplest case considered here. The maximum x and x' values at any point along s are defined by the emittance and the β and γ values as displayed in Eq. (4.3). The square root of ε therefore sets the scale factor A postulated previously, Eq. (4.1), and the dimensions of ε is length. The phase space ellipse in terms of $\varepsilon\alpha$, β, and γ is shown below in Figure 4.1 and has some boundary points labeled in that figure. The maximum values of x and dx/ds depend on the parameters β and γ, respectively. Note that 95% of all beam particles are contained within a 2σ boundary for a beam distributed as a Gaussian.

$$x_{\text{int}} = \sqrt{\varepsilon/\gamma}, \quad x_{\text{max}} = \sqrt{\varepsilon\beta} \sim \sigma_x$$

$$x'_{\text{int}} = \sqrt{\varepsilon/\beta}, \quad x'_{\text{max}} = \sqrt{\varepsilon\gamma} \sim \sigma_{x'} \quad (4.3)$$

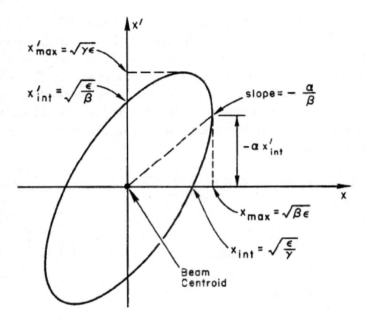

Figure 4.1:   The phase space ellipse in the (x, x') plane with area $\pi\varepsilon$, emittance ε. The maximum values of x and x' are shown as well as the intersections of the ellipse with the x and x' axes.

The ellipse boundary, shown in Eq. (4.4), can be alternatively formulated by using the actual occupation of a particular beam within the allowed phase space set by some limiting apertures. The area of the ellipse is $\varepsilon\pi$. For an emittance $\varepsilon \sim 12\pi * \text{mm} * \text{mrad}$ very roughly, for a beam pipe with a 2 cm radius, angles of about 01.9 mrad are accommodated. The envelope of the beam depends on the square root of $\varepsilon\beta$, which is largest typically near an F quadrupole and smallest near a D quadrupole. The Fermilab Main Ring, MR, will be used in examples that follow in order to have a uniform numerical set of parameters that the reader can become familiar with. However, the formalism applies to any repetitive focusing structure.

$$\varepsilon = \gamma x^2 + 2\alpha(xx') + \beta(x')^2 \qquad (4.4)$$

The ellipse boundary can be formulated in terms of a matrix of the lattice parameters contracted to a scalar emittance. This identity is easy to establish symbolically in Matlab as are the other definitions used below.

$$\gamma x^2 + 2\alpha x x' + \beta(x')^2 = \varepsilon$$

$$\begin{pmatrix} x & x' \end{pmatrix} \begin{pmatrix} \gamma & \alpha \\ \alpha & \beta \end{pmatrix} \begin{pmatrix} x \\ x' \end{pmatrix} = \varepsilon \qquad (4.5)$$

The r.m.s. averages of x^2, $(x')^2$ and the correlation xx' define an actual beam. The beam matrix, Σ, is given in Eq. (4.6). It is related to the lattice parameters α, β, and γ and the emittance ε and is a symmetric 2×2 real matrix defined by three parameters equivalent to α, β, and ε. It has determinant of 1, $|\Sigma| = 1$. The x, x' correlation is $\langle xx' \rangle = -\varepsilon\alpha$. The ellipse is shown in Figure 4.2 with specific points labeled by elements of the σ matrix, Eq. (4.6), and the emittance.

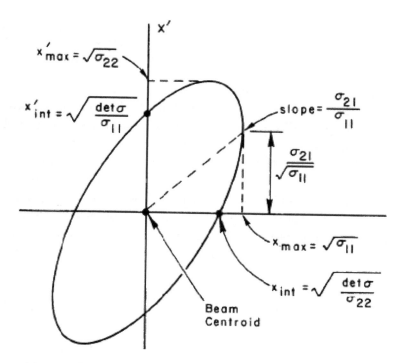

Figure 4.2:   Beam ellipse labeled by the alternative method of using the elements of the beam matrix where $\sigma = \varepsilon\Sigma$.

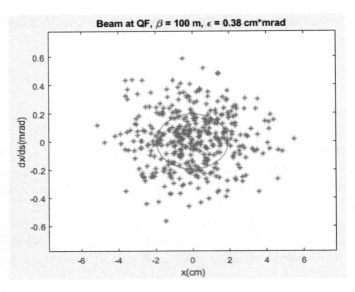

Figure 4.3: Plot of (x, x') for 400 Monte Carlo generated beam particles at a Q_F location with the maximum β of 100 m. The red ellipse boundary is for the 1 σ boundary.

A Monte Carlo generated ellipse using the "Gaus" algorithm from Section 3.12 to populate an ellipse at a Q_F location is generated using the script "Ellipse_Gaus" and the plot appears in Figure 4.3. The emittance is chosen to be 3.8×10^{-6} m * rad which will be used consistently. Some of the population falls outside the ellipse since only 68% of a Gaussian area is within 1 standard deviation and only 95% within 2. The ellipse is shown in red, while for 400 generated points, 20 are expected to fall outside the 95% boundary. Clearly, it is important to not lose a significant amount of beam at an aperture, so that such an aperture should be several "sigma" away from the beam.

$$\begin{pmatrix} \langle x^2 \rangle & \langle xx' \rangle \\ \langle xx' \rangle & \langle x'^2 \rangle \end{pmatrix} = \sigma = \varepsilon\Sigma = \varepsilon \begin{pmatrix} \beta & -\alpha \\ -\alpha & \gamma \end{pmatrix}, \quad \gamma = (1 + \alpha^2)\beta \ (4.6)$$

The beam matrix can be transformed from one location in s to another using the transfer matrix from an initial point i to a final point f, M, as shown in Eq. (4.7). The superscript T

indicates the matrix transpose. These matrix operations are easily performed, either symbolically or numerically, using the Matlab matrix utilities. This fact makes Matlab particularly useful in both beam and accelerator applications. Note that the properties of M are quite general in the formulations of the classical mechanics of conservative systems. The determinant $|M|$ is equal to 1, and the transpose is the inverse in this case.

$$
\begin{pmatrix} x \\ x' \end{pmatrix}_f = M \begin{pmatrix} x \\ x' \end{pmatrix}_i
$$

$$
\Sigma = \begin{pmatrix} \beta & -\alpha \\ -\alpha & \gamma \end{pmatrix}, \quad \Sigma_f = M \Sigma_i M^T
$$

(4.7)

These are two complementary ways to treat the properties of a beam, either using the α, β, μ lattice parameters or the beam parameters, $\langle x^2 \rangle$, $\langle x'^2 \rangle$ and $\langle xx' \rangle$. The invariant emittance in the two cases is found using the script "emit_abg_sig_M". The transfer matrix M is used to propagate the beam ellipse from location to location. The general transformations for the three lattice parameters in terms of the transfer matrix M is also shown, in Figure 4.4. It is not particularly illuminating however.

The emittance is $\langle x^2 \rangle / \beta$ for an on-momentum beam particle. With a limiting aperture of a at the location of the maximum value of β, usually an F quadrupole, the emittance is calculable knowing β. The emittance is a fundamental property which needs to be well measured. To lowest order it can be estimated using the calculation of the lattice parameters and the known limiting apertures. The actual emittance will normally be less than this first estimate and must be measured case by case. A few techniques used to measure the emittance will be mentioned later in the text.

If the beam energy changes it is useful to have an invariant emittance. Since the transverse momenta are approximately constant during acceleration while the longitudinal momenta increase, the value of dx/ds goes down. The invariant or normalized emittance has a factor (p/mc) added to compensate for the emittance reduction

```
>> emit_abg_sig_M

e =

g*x^2 + 2*a*x*xp + b*xp^2


gnew =

b*M12^2 - 2*a*M12*M22 + g*M22^2


bnew =

b*M11^2 - 2*a*M11*M12 + g*M12^2|


anew =

M12*(M12*a - M11*b) + M22*(M11*a - M12*g)
```

Figure 4.4: Output of the script "emit_abg_sig_M". The emittance in terms of the lattice parameters is found using Eq. (4.5) and then the transformed values of the lattice parameters are printed in terms of the initial parameters and the elements of the transfer matrix using Eq. (4.7).

due to acceleration, $\varepsilon_N \sim (\langle x^2 \rangle / \beta)(p/mc)$ where β is the lattice parameter and p is the beam momentum. As p increases the r.m.s. beam size decreases while the normalized emittance is unchanged.

Finally, for reference, the transformation to a position s which is not $s_o + C$, takes a fairly simple form when α is equal to zero, or when β takes a maximum or minimum value, $\alpha = 0$ and $\gamma = 1/\beta$. That occurs in an F or D quadrupole when β is a maximum or minimum, respectively. The phase advance from s_o to s is $\Delta\psi$ in Eq. (4.8). This can be a useful way to find x and x' at the center of either F or D quadrupoles. Note that $|M| = 1$ is particularly clear in this case.

$$
\begin{pmatrix} x \\ x' \end{pmatrix} = M(s|s_o) \begin{pmatrix} x_o \\ x_o \end{pmatrix}
$$

$$
= \begin{pmatrix} \sqrt{\beta/\beta_o}\cos\Delta\psi & \sqrt{\beta\beta_o}\sin\Delta\psi \\ -\sin\Delta\psi/\sqrt{\beta\beta_o} & \sqrt{\beta_o/\beta}\cos\Delta\psi \end{pmatrix} \begin{pmatrix} x_o \\ x_o' \end{pmatrix} \quad (4.8)
$$

For reference the general transformation is shown in Eq. (4.9). It too is not very transparent in regards to the physical meaning of the formalism. This general expression has a limiting case when $s = s_o + C$, where $\alpha = \alpha_o$, and $\beta = \beta_o$ which was shown previously in Eq. (4.2) and where, in that case, the phase advance is μ.

$$M(s|s_o) = \begin{pmatrix} \left[\sqrt{\beta/\beta_o}\right]\left[\cos(\Delta\psi\right) \\ + \alpha_o\sin(\Delta\psi)] & \sqrt{\beta\beta_o}\sin(\Delta\psi) \\ \\ [(\alpha_o - \alpha)\cos(\Delta\psi) \\ - (\alpha\alpha_o + 1)\sin(\Delta\psi)]/\sqrt{\beta\beta_o} & \sqrt{\beta_o/\beta}[\cos(\Delta\psi) \\ - \alpha\sin(\Delta\psi)] \end{pmatrix}$$

$$(4.9)$$

After this mass of mathematical formalism, we now turn to the basic building block of a repetitive accelerator structure, the FODO. This structure provides overall focusing for the circulating beam and normally also provides the dipole bending needed to create a circular path for the beam with circumference C.

4.2. FODO

A FODO is a quadrupole doublet, usually with dipoles at the "O" locations, in which the focal lengths of the F and D quadrupoles are equal. This is in contrast to the beam line doublet. For the FODO, F focuses vertically and defocuses horizontally, and D focuses horizontally and defocusses vertically. The result is a net, fairly gentle, focusing in both the x and y directions and only a single power bus is needed. This arrangement is a logical extrapolation of the beam line doublet to stable operation with many beam passages through a system of identical doublets. The goal is not that of a beam transport to focus at a point, but to transport the beam stably and without unnecessary elements. In the absence of insertions for injection, extraction, or collisions, an accelerator is a simple repetition of the basic FODO unit cell.

The FODO consists of 1/2 of an F quadrupole, bending magnets, a D quadrupole of the same focal length, a second set of dipole magnets, and finally 1/2 of an F quadrupole as shown in Figure 4.5.

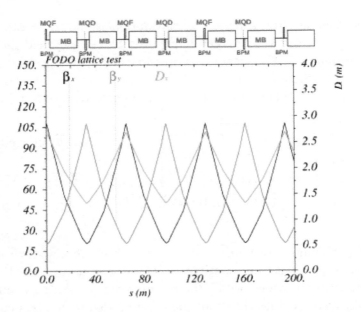

Figure 4.5: Plot of the β_x and β_y functions as well as the periodic dispersion, D, for three successive FODO.

The 1/2 quadrupole is chosen because the x excursion of the beam is largest there. The equal strength quadrupoles provide overall focusing in both x and y, while the dipoles define the momentum of the reference trajectory. Ideally, the quadrupoles and the dipoles would each need only a single common power bus. Typically the dipoles take up the bulk of the FODO real estate. A plot of the β_x and β_y functions and the periodic dispersion D (defined later) is shown below for several FODO operated together. The dipoles bend in the horizontal plane. It is assumed, in lowest order of approximation, that the x and y motions are independent. The location of many Beam Position Monitors (BPM) is notable, since they provide crucial diagnostic data on the actual behavior of the accelerator beam and would have a granularity sufficient to perform the required diagnostics.

The FODO as unit cell is explored in the script "FODO_Unit". Symbolic printout of the 2×2 transfer matrix for the focal length, $f_D = f_F = f$, and the d distance between Q_F and $Q_D = d$, is

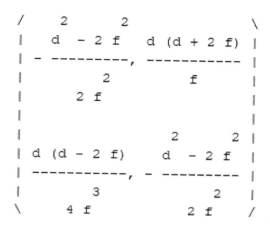

Figure 4.6: Transfer matrix in (x, x') for a FODO in thin lens approximation. The focal length of the quadrupoles is f, and d is the distance between the centers of the F and D quadrupoles. The y matrix is similar.

sent to the Command Window and displayed in Figure 4.6. The transfer matrix for the basic FODO is easily constructed from the matrices for quadrupoles and drift spaces (the dipoles have no effect on the reference trajectory) in the thin lens approximation which were shown previously in Section 2.5 and 2.6. More detailed solutions for finer grained behavior can be extracted from the appropriate thick lens transfer matrices, as was done in Figure 2.14 for a beam line doublet.

Treating the FODO as the repetitive unit and applying Eq. (4.2), the lattice parameters can be extracted as shown in Eq. (4.10), since they and the transfer matrix are alternate descriptions of the situation. The trace of M is equal to $2\cos\mu$ and $\cos\mu = 1 - (1/2)(d/f)^2$. Stability of the lattice requires $|\cos\mu| < 1$. There is net focusing in both x and y if $M(2, 1) < 0$ or $f > d/2$. The FODO phase advance is $\sin(\mu/2) = d/2f$.

$$\beta_o = M(1, 2)/\sin\mu, \quad \alpha_o = [M(1, 1) - \cos\mu]/\sin\mu, \quad \gamma_o = [1 + \alpha_o^2]/\beta_o$$
$$(4.10)$$

In the thin lens approximation, the maximum and minimum values of β in the FODO occur at the center of the F quadrupole and the D quadrupole respectively and have values defined by d and the phase advance as shown in Eq. (4.11). They are found using $M(1,2) = \beta \sin\mu = d(d+2f)/f$.

$$\beta_{\max} = 2d(1 + \sin\mu/2)/\sin\mu$$
$$\beta_{\min} = 2d(1 - \sin\mu/2)/\sin\mu$$

$$(4.11)$$

A concrete example of the parameters of a FODO is shown in the table of Figure 4.7 for the old Fermilab Main Ring, (MR). The emittance is taken to be $\varepsilon_{\mathrm{MR}} = 3.8\,\mathrm{mm} * \mathrm{mrad}$ as was the case for Figure 4.3. For the parameters of the MR FODO given in the table below, the phase advance is 70.91°. For a circular "accelerator" built of 96 unit cells of FODO, the total phase advance per turn, in units of 2π, or the tune Q, would be 18.91. The FODO dipoles are two strings, each bending by 33 mrad, for a total bend angle in 96 FODO of 6.34 rad $\sim 2\pi$. The tune has a "chromaticity" because the quadrupole focal length depends on momentum. For a 1% change in the momentum there is a 1% change in the FODO focal length, leading to a phase shift of 0.7°, or $\sim 1\%$. The dispersion due to off-momentum particles will be discussed in Section 4.3.

Phase advance (degrees) μ	70.91
Quadrupole length (m)	2.1
Distance between quadrupoles – d(m)	29.7
βmax (m)	99.4
βmin (m)	26.4
Quadrupole focal length (m)	25.6
Quadrupole gradient (T/m)	24.5

Figure 4.7: Parameters of the MR FODO which will be used consistently throughout the numerical examples in the text in order to provide continuity.

A simple circular "accelerator" would in general consist of N FODO where the reference trajectory is defined by the bending magnets which provide a circular circumference after traversal of the N unit cells. The "Q" of the accelerator is the number of oscillations made in the one turn traversal. The value of L is the length of the unit cell, $\sim$59.4 m in the MR example. The Q value need not be an integer since the lattice repeats, but the phase advance does not. The approximate average value of β in the FODO is estimated to be 48.5 m for the MR example. Using the actual MR value for R of $\sim$1 km, the radius of the "accelerator", leads to an estimate of $\langle \beta \rangle$ of 53 m. The situation is similar to that shown in Figure 4.5.

$$Q = N\mu/2\pi = (N/2\pi) \int_o^L ds/\beta(s) \sim NL/(2\pi \langle \beta \rangle)$$

$$\langle \beta \rangle = C/2\pi Q = R/Q, C = NL$$

(4.12)

The general FODO can be explored using the App "FODO_Unit". The specific result for the initial MR parameters is shown in Figure 4.8. Since all the parameters can be determined when f, d and ε are specified, many parameters appear numerically in the "EditFields". In addition a plot is made of the x and y as a function of z assuming a parallel beam input which shows the overall focusing. Another plot shows the behavior of β_x, β_y and the periodic dispersion D, in m, within the FODO similar to the plot of Figure 4.5. Finally the behavior of the ellipse is shown by making a "movie" starting in the Q_F center and going four steps in drift space then to the center and exit of Q_D, followed by four more drifts and ending back at the Q_F center. The values of β_x and D_x peak in Q_F, while for β_y the peak occurs in Q_D. The user can freely choose values for f and d using the two "Sliders" to create other FODO layouts. The dispersion, η, defined to be the change in x per dp/p as a function of s, is also shown using "EditField"s. This function is similar to that shown in Figure 4.5 and will be defined and discussed in the next section. Typically, in this text D is used for FODO matrix elements such as $M(1,3)$, while $\eta(s)$ is used for general accelerator dispersion, $dx(s)/(dp/p)$ as a function of s. The function $\eta(s)$ is periodic as are the other lattice parameters such as $\beta(s)$.

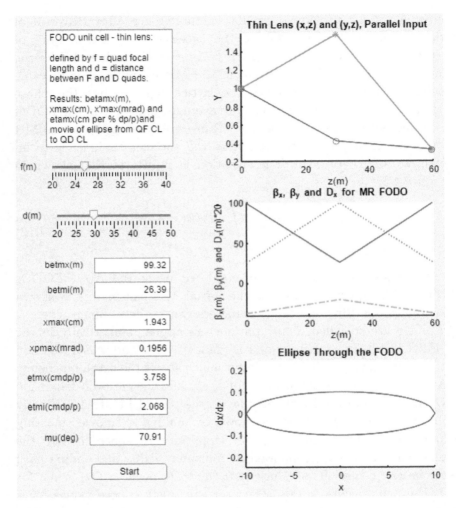

Figure 4.8:   Output of the script "FODO_Unit". The values of f and d are chosen by "Slider" which specifies the parameters of the FODO, except for dispersion. The FODO dipoles bend by a fixed 0.066 rad angle. The ellipse is shown in a movie going through a full FODO unit cell.

4.3.   Dispersion — δ

There is another dimension needed to characterize a real beam. Any stable beam contains a range of momenta dp close to the design momentum p_o. So far, the limits on this range have been ignored.

The focal length of a quadrupole depends on the particle momentum, $f = p/[Lqe(dB/dr)]$, Eq. (2.16), as does the bend angle of the particle momentum in a dipole, $\rho = p/(eqB)$, $\phi_B = L/\rho$, Eq. (2.10). Beams in accelerators have a finite acceptance for momenta different from the momentum on the reference trajectory, $dp = p - p_o$.

The quadrupole effects on off-momentum particles are largely ignored in this text and are assumed to be compensated for by the judicious use of sextupole elements leaving only small chromatic effects due to the quadrupoles. The tune shift due to quadrupole focal length changes without correction is estimated in Eq. (4.13). The tune shift is defined for a full "turn" with $s = C$, and is N times the tune shift for the FODO with $N = 96$ for the MR example. The shift is the integral over ds from 0 to C of $(-1/4\pi)$ times $\beta(s)k(s)$ times dp/p. It is evaluated accurately later in Section 4.10. For the MR FODO example the estimate is, $dQ \sim 0.22$ for a 1% dp/p which is a dQ/Q of 1.2%. The proportionality between dQ and dp/p is the dimensionless parameter called chromaticity. For the MR example it is $\sim -21.7 \sim Q$. Another numerical order-of-magnitude estimate for the FODO example used in the text appears below Eq. 4.11 in agreement with this estimate. Strong FODO focusing increases the tune shift. There is a tune bandwidth due to the dp/p distribution which may approach a destabilizing resonance and must be corrected for in that case.

$$dQ \sim -(N/4\pi)(\beta_F - \beta_D)(1/f)(dp/p) \qquad (4.13)$$

The major effect of dispersion on the accelerator comes from the dipoles. The off-momentum particle in a dipole is no longer on the reference trajectory, going further if of higher momentum and having an exit angle not that of the reference trajectory, but smaller. These effects are displayed in Figure 4.9. At the lowest order for a dipole a 3×3 matrix can account for these effects. The position offset and the angle offset for a dipole of length L have already been seen in the previous discussion of trajectories in a dipole in Chapter 2.

The matrix formulation has been for the reference momentum so far. In order to treat off-momenta, the dimensionality of the matrices must increase to 3, for the x coordinate in which the dipole bends are

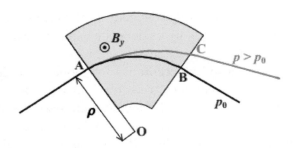

Figure 4.9:   Trajectory of an off-momentum particle in a sector dipole, where the entrance and exit angles are perpendicular for the reference trajectory. For a high-momentum particle the path length is longer and the bend angle is less.

assumed to occur. The lowest order dipole (3×3) transfer matrix, for a dipole of length L, has the added terms, $M(1,3)$ and $M(2,3)$ while the previous (2×2) elements of the matrix are unchanged and the 2×2 dipole acts as a drift space. The dipole matrix appears in Eq. (4.14). The change in x due to δ is $\phi_B L/2$ while the change in angle is ϕ_B. These changes have previously been mentioned in Eq. (2.11).

$$M_B \sim \begin{pmatrix} 1 & L & \phi_B L/2 \\ 0 & 1 & \phi_B \\ 0 & 0 & 1 \end{pmatrix}, \quad \begin{pmatrix} x \\ dx/ds = x' \\ dp/p = \delta \end{pmatrix} \qquad (4.14)$$

The FODO has so far been treated as having two drift spaces between the $Q_F/2$ and the Q_D. In fact, the space is filled with dipoles, as seen in Figure 4.5. The expanded matrices for the thin lens FODO has elements $M(1,3)$ and $M(2,3)$ generated in the script "FODO_Dispersion" and shown in Figure 4.10. Since the 2×2 sub-matrices for x and dx/ds are unchanged, the lattice parameters do not change in this level of approximation.

For the thin lens FODO the periodic dispersion, D, is proportional to $d\phi_B$ while D' is to ϕ_B alone times dimensionless functions

```
>> FODO_Dispersion
MT Transfer Matrix for the FODO, M(1,3), M(2,3)
d phi (d + 4 f)
---------------|
      2 f


     2                 2
 phi (d  + 2 d f - 8 f )
 - ----------------------
            2
          4 f
```

$$(4.15)$$

Figure 4.10: The D, $M(1,3)$, and D', $M(2,3)$, functions for the thin lens FODO. The length of each of two sets of dipoles in the FODO is d and each set bends by ϕ_B.

of the dipole length divided by the quadrupole focal length.

$$D = d\phi_B(2 + d/2f)$$
$$D' = \phi_B[2 - (d/2f) - (d/2f)^2]$$

$$(4.16)$$

The dispersion, η, is defined to be the change in x per unit of $\delta = dp/p$ at a point in the lattice, $\eta(s) = dx/d\delta$. It is a function of s, has units of length, and is periodic. The transfer matrix eigenvector is a column vector with elements η, η' and 1 for x, dx/ds and dp/p. Since η is periodic, $\eta = \eta \cos\mu + \eta'\beta \sin\mu + D$. Since η' vanishes at the quadrupole centers, Figure 4.5, $\eta = -D/(1 - \cos\mu)$. The generalized case is shown in Eq. 4.17. The values for the MR FODO were shown in Figure 4.8. A useful estimator for dispersion is that $\eta_x(s) \sim \beta_x(s)/Q_x$ which estimates a MR FODO maximum value 5.25 cm for 1% δ and a 1.4-cm minimum value. These estimates are certainly a useful approximate starting point.

$$\eta_{\max} = (d\phi_B)[1 + (1/2)\sin(\mu/2)]/\sin^2(\mu/2)$$
$$\eta_{\min} = (d\phi_B)[1 - (1/2)\sin(\mu/2)]/\sin^2(\mu/2)$$

$$(4.17)$$

The beam ellipse standard deviation has an added term due to dispersion, $\sigma_x^2 = \varepsilon_x\beta_x(s) + [\eta(s)\delta]^2$, which increases the size of the beam in x. The beam size is set by the lattice functions and the emittance with a second term, added in quadrature, which is proportional to dp/p. The dispersion is largest in the F quadrupole and smallest in the D quadrupole as seen previously in the plots in Figures 4.5 and 4.8.

The Fermilab MR FODO has two strings of dipoles with a 33 mrad bend each, for a total 66 mrad bend angle per FODO. With 96 FODO the full bend is $1.008 \times 2\pi$ approximating the periodic circular ring. The maximum and minimum values for a MR FODO β and η were shown in Figure 4.8. Taking an emittance of 3.8×10^{-6} m $*$ rad, the FODO ellipse maximum x is 2 cm and the maximum dx/ds is 0.2 mrad. The maximum η value is 3.8 cm for 1% dp/p, larger than the on-momentum ellipse value, while the minimum is 2.1 cm. Momenta with 1% deviations from the reference momentum will likely not be captured into a stable beam within a typical beam-pipe radius since several r.m.s. values of aperture are needed for safe operation. Indeed, a 0.1% dp/p is more likely as will be seen in the discussion the momentum acceptance for stable acceleration.

4.4. Acceleration — y, $\Delta\varepsilon$, $\Delta\phi$

The temporal properties of an accelerated beam of particles are defined by the accelerating field. Typically, a radio frequency (r.f.) electric field is used to accelerate a charged particle beam since all particles start essentially from being at rest. This results in a restricted region of time when the particles are stably accelerated. The variables here are the spread in the energy, $\Delta\varepsilon$, and the r.f. phase, $\Delta\phi$, with respect to the "synchronous" particles, on the orbit which are stably captured. This longitudinal phase space is the area in energy, and r.f. phase where stable orbits are possible. Off-momentum here is in the longitudinal direction. The beam's rotation frequency is ω_o, the rotation period is τ, the circumference is C, the radius of curvature is ρ and the r.f. accelerating frequency is ω_{rf}. In Eq. (4.18) the parameter h is called the harmonic number and is not the Planck constant.

$$\omega_o = 2\pi/\tau = 2\pi\nu/C = \nu/\rho \sim c/\rho$$
$$\omega_{rf} = h\omega_o$$

$$(4.18)$$

There are two effects for an off-momentum particle, the change in circumference, dC, and the change in velocity, dv as mentioned previously in the discussion of dispersion. A high-momentum particle must take a longer path length in the dipoles, Figure 4.9, but it will also have a higher velocity than the reference momentum. A compaction parameter, γ_t, is specific to a given accelerator and defines when one effect dominates over the other. The change in terms of the SR γ factor of the beam particle is defined to occur at "transition" shown in Eq. (4.19) when the two competing effects cancel.

$$1/\gamma_t^2 = (dC/C)/(dp/p) \qquad (4.19)$$

The change in period is sometimes defined to be the "dispersion", S, scaled by dp/p. This other definition, also called "dispersion", is dimensionless. It is sometimes referred to as the slip factor, S, which it will be in this text to avoid confusion with the dispersion η. In the previous lattice definition for the FODO D had dimensions of length and the dispersion $\eta(s)$ also had units of length but scaled by dp/p.

Far above transition $S \sim 1/\gamma_t^2$ and is positive while it is negative below transition. For the FNAL MR, $\gamma_t \sim Q_x$, and is ~ 18.7 at 150 GeV. Since the S factor is the integral around C of $\eta(s)/\rho$ over ds/C another useful estimate is that $S \sim \langle \eta \rangle / R$ which for the MR, with $R \sim 1$ km and $\langle \eta \rangle = 2.9$ m, is $\sim 0.0029 \sim 1/(18.7)^2$.

$$\tau = C/v$$
$$d\tau/\tau = dC/C - dv/v = (1/\gamma_t^2 - 1/\gamma^2)(dp/p)$$
$$= S(dp/p) \qquad (4.20)$$

There are two aspects of the allowed spread in momentum for an accelerator. First, there is the possible range of momenta where stable orbits can be captured, and which can then be accelerated. That is called the r.f. "bucket". The second is the actual range of momenta of particles in the accelerator, called the "bunch". Units for the bunch are often taken to be eVs. There is a range of r.f. phase, ψ, over which a particle may be captured into a stable orbit with a synchrotron frequency ω_{syn} having an s position which oscillates, due to the restoring force of the r.f. acceleration. A phase difference ϕ exists between the synchronous orbit, $\psi_s = \phi_s$, and the off-momentum orbit. The energy difference between a particle of that phase and the synchronous particle is defined to be $\Delta\varepsilon$. The mean energy gain per turn is $eV_o \sin\phi_s$ where V_o is the amplitude of the r.f. voltage.

$$d\psi = \omega_{rf}(\tau + d\tau), \omega_{rf} = h\omega_o$$

$$\phi = \psi - \omega_{rf}\tau$$

$$d\phi = \omega_{rf}\tau(d\tau/\tau) = [\omega_{rf}\tau S(dp/p)] = [\omega_{rf}\tau S d\varepsilon]/(\beta^2\varepsilon) \tag{4.21}$$

$$\Delta\varepsilon = \varepsilon - \varepsilon_s = qeV_o[\sin\phi - \sin\phi_s]$$

The intermediate step in Eq. (4.21) uses the result for (dp/p) in terms of $(d\varepsilon/\varepsilon)$. This is easily found using a small Matlab script snippet, Figure 4.11, to be $1/\beta^2$ which is printed in the Command Window by the script.

```
% SR derivatives
%
syms   e p g b r
g = 1 ./sqrt(1-b^2); e = g; p = b*g;
dp = diff(p); de = diff(e);
r = simplify((dp/p)/(de/e))
```

Figure 4.11: Matlab script snippet to find dp/p and $d\varepsilon/\varepsilon$.

The r.f. phases where the voltage is increasing with time are stable below transition, while the decreasing phases are stable above transition, at high energies. The focus in this text is well above transition where $d\tau/\tau$ is simply proportional to dp/p with a constant slip factor. That happens because at high energies all β approach 1 so that the higher momentum particles have to go farther and arrive late. As shown in Figure 4.12, those particles, above transition, receive less energy gain. Those that arrive early receive more energy gain. In both cases the change is toward stability and oscillatory motion is expected. There is a region of phase where the r.f. restoring force gives phase stability, the bucket. The maximum stable range of phase is approximately π or half of an r.f. period. It depends on the synchronous phase and the value of dp/p.

Instead of time, t, as a variable the number of turns, n, can be used with $dn = dt/\tau$. Given the slow speed of the synchrotron

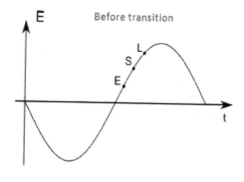

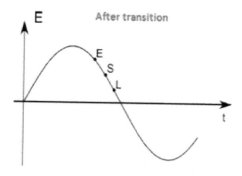

Figure 4.12: Phase stability above and below transition. Above transition, the dominant effect is that higher energy particles have a longer path and arrive late, L. In that case, particles that arrive early, E, receive more energy than those that arrive late so that there is a restoring force for both E and L particles.

oscillation, n can be considered a continuous variable. That assumption allows for the phase change, $d\phi/dn$, to be expressed as a function of $\Delta\varepsilon$ while $d(\Delta\varepsilon)/dn$ is a function of the phase ϕ. In this text the time variable will be used exclusively, although conversion from t to n is simple and some texts use n. For small values of the phase difference, a second-order differential equations leading to simple harmonic motion for the phase as a function of t is obtained. Since the rotation period τ is so much shorter than the synchrotron oscillation period, the variable can be switched from τ to dt without a problem. The equations shown in Eq. (4.21) can then be combined into a second-order differential equation in time for the phase ϕ as shown in Eq. (4.22). The restoring force arises when the phase differs from the synchronous phase.

$$d\phi/dt = (S\omega_{rf}/\varepsilon_s\beta^2)\Delta\varepsilon$$

$$d(\Delta\varepsilon)/dt = (qeV_o/\tau)[\sin\phi - \sin\phi_s] \qquad (4.22)$$

$$d^2\phi/d^2t = (S\omega_{rf}/\varepsilon_s\beta^2)(qeV_o/\tau)[\sin\phi - \sin\phi_s]$$

Subtracting out the synchronous phase, $\Delta\phi = \phi - \phi_s$, and assuming small phase deviations from the synchronous orbit, the approximate result for the synchrotron oscillation in terms of time is shown in Eq. (4.23) for operation far above transition. A synchrotron tune, Q_s, can be defined relative to the accelerator period.

$$d^2\Delta\phi/d^2t = \omega_{syn}^2\Delta\phi$$

$$\omega_{syn} = \sqrt{-S\omega_{rf}\omega_o(qeV_o\cos\phi_s)/2\pi\varepsilon_s\beta^2} \qquad (4.23)$$

$$Q_s = \omega_{syn}/\omega_o = (1/\beta)\sqrt{-Sh(qeV_o\cos\phi_s)/2\pi\varepsilon_s}$$

If V_o and ε_s are constant, then a constant of motion, analogous to energy for a harmonic oscillator, can be defined. The constant has a "kinetic" contribution proportional the $(d\phi/dt)^2$ and a "potential" contribution proportional to qeV_o. The invariant can be defined alternatively using either ϕ or $\Delta\varepsilon$ in Eq. (4.24) by using the relationship between $d\phi/dt$ and $\Delta\varepsilon$. It is a useful constant of motion.

$$1/2(d\phi/dt)^2 + [S\omega_{rf}(qeV_o)/\varepsilon_s\beta^2\tau][\cos\phi - \sin\phi_s]$$

$$(d\phi/dt) = [S\omega_{rf}/\varepsilon_s\beta^2]\Delta\varepsilon \qquad (4.24)$$

$$\Delta\varepsilon^2 + (2qeV_o\varepsilon_s\beta^2)/S\omega_{rf}\tau][\cos\phi - \sin\phi_s]$$

A specific example is the Fermilab MR. The r.f. has a harmonic number, h of 1113. The rotation frequency at $\beta = 1$ is 0.3 MHz while the r.f. is 333 MHz. A maximum energy gain per turn is 2 MV or 95 GeV/s. The approximate maximum value of $y = \Delta\varepsilon/\omega_{rf}$ is 0.734 eVs. The approximate synchrotron frequency is 780 Hz, much less than the rotation frequency. It takes about 385 turns for a particle to perform a synchrotron oscillation justifying the approximations used above a posteriori. The momentum acceptance depends on the acceleration and is approximately 0.32% FWHM at 150 GeV. In Eq. (4.25) the radius is R, or approximately ρ, and cp is the energy in $eV \sim \varepsilon_s$. The expression for $y_{\max}$ holds for UR motion and uses the results of Figure (4.11), $d\varepsilon/\varepsilon \sim dp/p$.

$$y_{\max} = \Delta\varepsilon_{\max}/\omega_{rf} \sim [R(cp)/(hc)](dp/p)$$

$$dp/p = \delta \sim \pm(1/cp_o)\sqrt{\varepsilon_s(2qeV_o\cos\phi_s)/(\pi hS)} \qquad (4.25)$$

$$\delta_{\max} = \sqrt{2qeV_o\cos\phi_s/(\pi hS\varepsilon_s)}$$

An App, "Bucket_MR_rf", is available to vary the parameters for the FNAL MR as a numerical example after all the mathematical formulae. The constants for the differential equations are, $d(\Delta\phi)/dt = Ay$, $dy/dt = B[\sin(\Delta\phi)\sin\phi_s]$, $A = (hc/R)^2(S/\varepsilon_s) = 2125/\mathrm{eVs}^2$ and $B = qeV_o/2\pi h = 286\,\mathrm{eV}$ assuming UR operation. In those terms, $\omega_{\mathrm{syn}} = \sqrt{AB} = 780\,\mathrm{Hz}$ and $y_{\max} = \pi\sqrt{B/A} = 0.734\,\mathrm{eVs}$. The initial value of y and the synchronous phase are chosen by "Slider". The full nonlinear equations of motion are solved using the MATLAB utility "ode45". The maximum and minimum values of y and ϕ and the momentum acceptance are displayed using "EditFields". Plots of y and ϕ as a function of time are displayed as well as the contour of allowable values. The user can also explore the regions of instability where particles would be lost from the r.f. bunch as a function of both synchronous phase and momentum acceptance. The physical bunch length corresponds to an r.f. phase acceptance of at most π. An output of the App appears in Figure 4.13 in a case where the solution is stable and harmonic. In this specific case the phase range is 0.74π.

For the MR the r.f. frequency is $f_{rf} = 53\,\mathrm{MHz}$ with a wavelength of 5.66 m and a period of 18.9 ns. The maximum bunch with $d\phi \sim \pi$. may be at most 9.4 ns long or 2.8 m. The bunch length in general is proportional to the momentum acceptance of the bunch, $\sigma_z \sim (cS/Q_s\omega_o)\sigma_\delta$. For the MR with $S = (1/18.7)^2$, $Q_s\omega_o = 780\,\mathrm{Hz}$ and with a dp/p value of 0.10%, the bunch length is 1.1 m. Bunches with smaller momentum spread will have smaller bunch sizes.

As a perhaps welcome change from the math, consider a simple schematic example of the measurement of the bunch length.

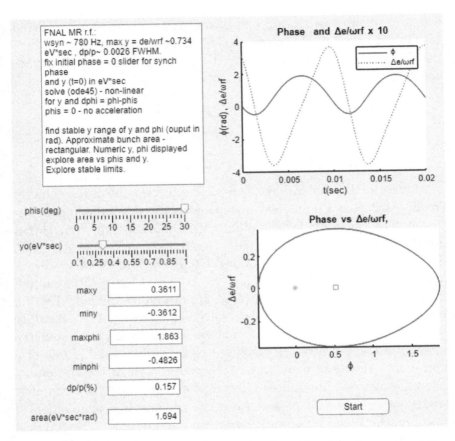

Figure 4.13:   Output for the App "Bucket_MR_rf". The values of ϕ_s and y_o define the bucket boundaries of y_{max}, y_{min}, ϕ_{max} and ϕ_{min} and the accepted momentum fraction. It is instructive to use the Sliders to explore the stability regions for phase and energy.

The bunch length can be measured using very fast lasers, by scanning in z, by looking at a scattered beam, or by displaying scattered laser photons. The bunch length depends linearly on the momentum acceptance, dp/p, so that a measurement of the bunch length determines dp/p. Alternatively, a measurement of the synchrotron frequency dependence on the r.f. voltage can be used to extract γ_t. Details will be given later for specific examples. For now, a first qualitative example of how to measure the bunch

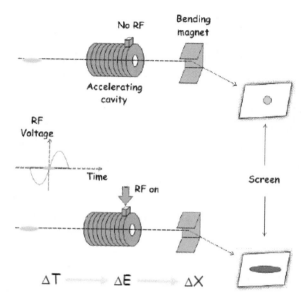

Figure 4.14: Schematic representation of the use of a special r.f. cavity to measure the bunch length in an accelerator.

length assumes there is a special r.f. cavity in the accelerator lattice.

As shown in Figure 4.14, with the help of a cavity and a bending magnet the beam is sent to a screen. With the cavity turned on and phased properly the energy gain or loss from the r.f. depends on the bunch location in z. The bending magnet dispersion then separates the different beam z into different locations on the screen depending on the time of the bunch at the location of the r.f. cavity. This is a destructive measurement however, and other methods must be used in order to preserve the beam that is being measured for its bunch length properties.

4.5. Electric Fields at $\beta \sim 1$

For NR motion it was shown in Eq. (3.15) that the energy deposited by a charged particle was a constant divided by β^2. The argument was then made that all particles deposit a fixed amount of energy, the minimum ionizing particle or mip, at UR velocities when

β approaches 1 independent of energy. It is timely to explore that assertion a bit more closely as regards the time development of the field. The electric field of a moving charge, with impact parameter b with respect to an observer at rest is examined using the App "E_Moving_q" as shown in Figure 4.15.

The NR electric force is $(qe)E_o = (qe)^2/(4\pi\varepsilon_o b^2)$ and acts over a time $\sim b/v$, so that the momentum impulse seen by the observer is $\Delta p_x = (qe)E_o\Delta t$. The effect of SR is to increase the field by a factor γ, but the relevant time shrinks by a factor γ, leaving the momentum impulse constant. In the UR regime, the $E_x(t)$ pulse is almost a "delta function" acting strongly but only over a very short time. The situation is like that of time impulsive noise shown previously which

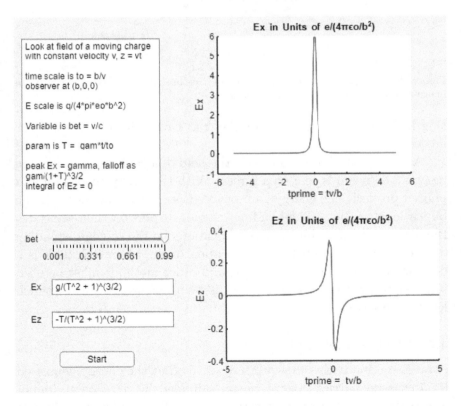

Figure 4.15: Output of the App "E_Moving_q". There is a "movie" of E_x and E_z in time followed by plots of both field components as a function of time. The formulae for E_x and E_z are shown symbolically with "EditFields".

occurs at all frequencies and thus had to be bandwidth limited in silicon detectors. The user chooses β by "Slider" and views a "movie" of the E vector as a function of time. After the movie plots of $E_x(t)$ and $E_z(t)$ are created. The field components are shown symbolically using "EditFields". Changing the "Slider" values of β is strongly encouraged.

4.6. Fourier Series and FFT

Since an accelerator has periodicity, the use of Fourier tools will be appropriate in the study of accelerator performance. Any periodic function can be represented by a series of harmonic functions. The Fourier series uses cos and sin functions. The series terms are shown in Eq. (4.26). The expansion is over one period, τ, of the function $y(u)$. The constant coefficient, a_o, is just the average value of $y(u)$ over the period. The other coefficients are calculable, either symbolically or numerically. Examples are shown using the App "Fourier_Series_App" of a few functions which are chosen using the "DropDown" menu. A square wave function is shown in Figure 4.16, where the function, the series and the series coefficients are displayed. The movie with increasing number of terms shows how the series more closely approximates the function. The user is encouraged to look at the functions that are supplied and to see how well the function is represented as the number of terms in the series varies up to 6. The constant term, a_o, is not displayed only the first five terms a_j.

$$y = a_o/2 + \sum_k [a_k \cos(k\omega u) + b_k \sin(k\omega u)]$$

$$a_k = 2 \int y(u) \cos(2\pi k u) du, \quad b_k = 2 \int y(u) \sin(2\pi k u) du, \quad (4.26)$$

$$\omega = 2\pi/\tau, u = t/\tau$$

The Fourier transform, F, is appropriate for a pulse train over a long period of time not just a single period. It is defined such that $F(\omega)$ is the integral over all time, t, of the function $f(t)e^{(j\omega t)}$, and is a function of frequency, ω, where j is the imaginary number $\sqrt{-1}$. There are a few calculable FT of simple functions. For example, the

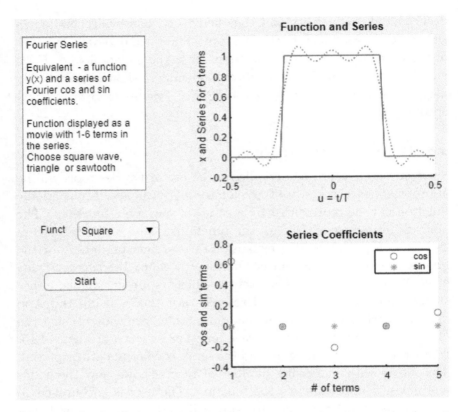

Figure 4.16: Output of the "Fourier_Series_App" for a specific "DropDown" choice of function.

FT of a delta function is uniform over all frequencies as was already mentioned in the discussion of noise in semiconductor detectors. A few of these simple FT are shown in Figure 4.17. One notable FT is that the FT of a Gaussian with an r.m.s. σ is a Gaussian with r.m.s. proportional to $1/\sigma$.

A few other examples of FT are displayed in the App "Fourier_Transform". They are chosen from a "DropDown" menu, but the user can try their own input as an option. The "EditField" $fn(t)$ is a user supplied function and $Fn(w)$ is the result, if it can be found by Matlab. The menu results for $F(\omega)$ are displayed symbolically. The script uses the Matlab utility "fourier" which may or may not find a solution for any particular user supplied function.

$x(t)$	$\hat{X}(f)$
$\delta(t)$	1
1	$\delta(f)$
$\cos(2\pi f_0 t)$	$\dfrac{\delta(f - f_0) + \delta(f + f_0)}{2}$
$\sin(2\pi f_0 t)$	$\dfrac{\delta(f - f_0) - \delta(f + f_0)}{2j}$
$\displaystyle\sum_{n=-\infty}^{\infty} \delta(t - nT)$	$\dfrac{1}{T} \displaystyle\sum_{k=-\infty}^{\infty} \delta\!\left(f - \dfrac{k}{T}\right)$
$\alpha e^{-\alpha t} u(t)$	$\dfrac{1}{\frac{j2\pi f}{\alpha}+1}$, $\alpha > 0$
$\mathrm{rect}(\alpha x)$	$\dfrac{1}{\lvert a \rvert}\,\mathrm{sinc}\!\left(\dfrac{f}{a}\right)$
$\mathrm{sinc}(\alpha x)$	$\dfrac{1}{\lvert a \rvert}\,\mathrm{rect}\!\left(\dfrac{f}{a}\right)$
$e^{-\alpha x^2}$	$\sqrt{\dfrac{\pi}{\alpha}}\,e^{-\frac{(\pi f)^2}{\alpha}}$

Figure 4.17:   A few of the simple FT for harmonic, δ, and Gaussian functions. Note that f and not ω is used in this figure.

A sample result appears in Figure 4.18 for a Gaussian distribution. Other functions such as "dirac$(t+2)$" and "heaviside$(t+2)$" are also solvable, for example. In regards to notation, $j = \sqrt{-1}$ in this text, while i is the particle current. In Matlab script li is $\sqrt{-1}$.

A tool of more practical use is the discrete Fourier transform or the fast Fourier transform (FFT). It is applied to a sequence of N time samples. The FFT is expanded in a series of N harmonic terms. This series can be inverted and for real functions the expansion is in terms of N cos and sin functions, analogous to the Fourier coefficients shown in Eq. (4.26). These operations are available in Matlab using the utilities "fft", "ifft" to find the FFT and the inverse FFT of

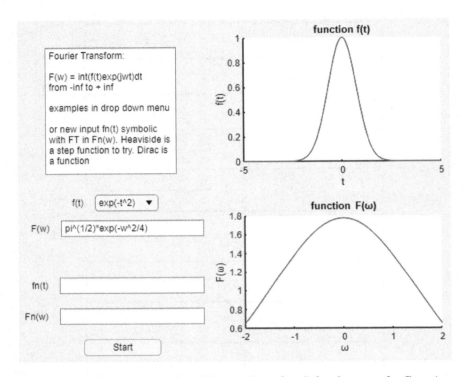

Figure 4.18:   Output of the App "Fourier_Transform" for the case of a Gaussian function.

a series of data points, and the utility "abs" which is used to find the power as a function of frequency ω. The FFT operation is on a discrete series of data points, not a function. In this text only the power is explored and not the phase information that is also available using the "fft" utility.

A script "FFT_sunspots" is supplied to introduce the Matlab FFT utilities. In fact, Matlab supplies 288 years of data on sunspot activity. In the script the data is displayed and the FFT is found. Then the FFT power is found as a function of frequency or period and the maximum power is computed. The sunspot cycle has a period of 11.08 years. The plots of sunspot activity in both time and frequency are displayed in Figure 4.19. The two descriptions are, indeed, complementary. The user can look at the script with the "comment" lines in order to become familiar with the "fft"

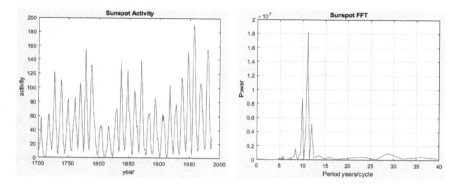

Figure 4.19:   Annual data for sunspot activity for 288 years and the FFT of the data, indicating an approximate 11 year cycle.

utility. Alternatively, there is Matlab documentation available using the "Search Documentation" window.

4.7. Induced Charge — Beam Pipe

The beam in an accelerator is performing betatron oscillations in x and y and synchrotron oscillations in z. It also contains bulk properties like total charge and perhaps bunch charge. The instrumentation that is deployed acts to measure these properties of the accelerator beam. For a beam of charge qe and total number of particles N revolving with frequency f the beam current is $I_b = qeNf$. The beam has high impedance and is effectively a perfect current source as shown in Eq. (4.27). For n charges, each qe, per unit length moving with velocity βc with an external applied voltage V, the impedance is shown in Eq. (4.27). The factor $d\varepsilon/d\beta$ is easily found symbolically using the utility "diff". The γ^3 factor says that as β approaches 1 it is hard to change it much, meaning that it is hard to change the beam current much. For example, a 400 GeV proton beam with a current of 1 A has an impedance of $4 \times 10^{13}\Omega$.

$$I_b = qen\beta c, Z_b = (dV/dI_b)$$
$$\varepsilon = \gamma mc^2, d\varepsilon/d\beta = -mc^2\beta\gamma^3, d\varepsilon = qedV \qquad (4.27)$$
$$Z_b = (mc^2\beta^2\gamma^3)/(qeI_b)$$

The electric field of a moving charge becomes an approximate delta function in time at high energies and the field direction is transverse, or a TE field. In that case all frequencies of the beam are equally probable. As with the discussion of PWC cathode strips, the beam induces a distributed charge on the conducting beam pipe. The beam linear current density at the beam pipe is $j(t) = I_b(t)/(2\pi a)$ where a is the beam pipe radius so that $j(t) = qeN(\beta c/2\pi a)$, and N is the total particle number in a bunch or the total in the beam. As discussed previously, the induced charge on the beam pipe resides on the surface since the skin depth at high frequencies is small. For example, the skin depth δ for Cu is $6.6\,\text{cm}/\sqrt{(f(Hz))}$, or $21\,\mu\text{m}$ at $10\,\text{MHz}$. As before, the method of image charges can be employed to explore the current density as a function of the location of the beam. Note that j here is the linear current density and not the imaginary number, $\sqrt{-1}$.

Using the induced charge on the beam pipe is a nondestructive method to determine the properties of the beam. The induced charge is treated in analogy to the PWC where the charge appears instantaneously at a point as a problem in electrostatics. The solution can be obtained using the method of image charges, but now for a cylindrical beam pipe as a function of the (x,y) location inside the beam pipe. The situation is approximated as a two-dimensional problem as the electric field is assumed to have zero extent in z. For a charge qe with a radius r inside the pipe of radius a the image charge is along the radius to the charge with image charge $q_i = -qe(a/r)$ and location $r_i = a^2/r$. The image charge induces a linear current density j on the wall at (a,Φ) due to the beam charge at (r,ϕ) as shown in Eq. (4.28). Assuming a beam position monitor (BPM) pickup exists covering an angle α around the x axis, j can be integrated over to find the total induced current I. To lowest order the induced current is proportional to the beam current and the BPM extent in azimuthal angle.

$$j(a, \Phi) = -(I_b/2\pi a)[(a^2 - r^2)/(a^2 + r^2 - 2ra\cos(\Phi - \phi))]$$

$$I = a \int_{-\alpha/2}^{\alpha/2} j(a, \Phi)d\Phi \sim (I_b/2\pi)\alpha$$

(4.28)

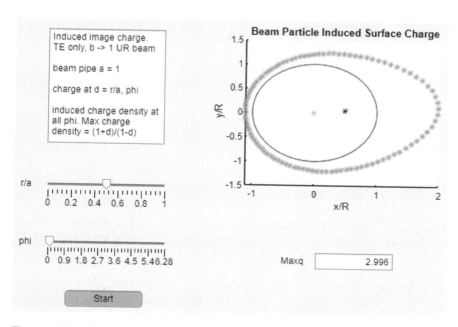

Figure 4.20:   Output of the App "Induced_Charge_TE". The black point is the beam, the green dot the center of the pipe, shown in blue, and the red dots are the induced charge.

The induced charge is displayed as a function of r/a and ϕ using the App "Induced_Charge_TE" with a specific result shown in Figure 4.20. Using the "Sliders" the user can move the beam charge anywhere inside the beam pipe and see the resulting induced surface charge.

Consider a few of the complications needed to read out the beam pickup with image charge Q, $I \sim dQ/dt$. The current is, Eq. (4.28), $\sim (I_b/2\pi)\alpha$, so for full azimuthal coverage I would be the beam current. With capacitive coupling into a resistive load, R, the impedance is Z, and the voltage is $V(\omega) = Z(\omega)I_b$. Typical circuitry is shown in Figure 4.21. For a pickup of area A, at radius a away from a centered beam the image charge, treated as a current source, is capacitively coupled, C, to a load resistor R. The impedance is $A/(2\pi a C \beta c)$ times the ω dependent effects of the R and C. The voltage is then frequency limited to $\omega > 1/\tau_c$, with $\tau_c = RC$, as

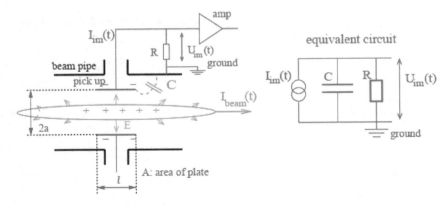

Figure 4.21: Schematic of the configuration of a beam pickup showing the beam idealized as a current source, with stray capacitance, C, and the load resistance R.

seen in Eq. (4.29). For high frequencies, $\omega \gg 1/\tau_c$, $V(t) \sim I_b(t)$ but for low frequencies, $\omega \ll 1/\tau_c$, the voltage is proportional to the time derivative of the current, $V(t) \sim dI_b(t)/dt$, which means there is poor response for long bunch lengths. The RC circuit is a "high-pass" filter. For beam current $I_b(t) \sim qeN\delta(0)$, $I_b(\omega)$ is uniform. For a Gaussian bunch length in t, σ, the frequency spectrum is $I_b(\omega) \sim qeN \exp(-(\omega\sigma)^2/2)$ as seen previously in Fig. 4.17 and $\sigma_\omega^2 = 1/\sigma^2$. If $V(t) \sim dI_b(t)/dt \sim qeN[t\exp(-t^2/(2\sigma^2))]/\sigma^3$, the voltage is a differentiated bipolar pulse.

$$V(\omega) = (1/\beta c)(1/C)(A/2\pi a)[j\omega\tau_c/(1 + j\omega\tau_c)]I_b(t) \qquad (4.29)$$

The bunch length can vary widely, from 10 ps (e linac), to 100 ps (p linac) up to 10 ns (p collider). Previously, in Section 4.4 the FNAL MR bunch length was estimated to be 1.1 m or 3.7 ns. In the case of a finite γ the induced current is not a transverse delta function but has a typical time spread $= a/(\gamma v)$ as shown in Section 4.5, with a frequency scale, $\omega \sim (\gamma\beta c)/a$. This limitation typically applies to linacs and other lower velocity machines which have induced charges more spread out in time as discussed later in Section 4.9.

4.8. Beam Position Monitor — x, y

A BPM is fashioned typically by the use of two pickup electrodes. The electrodes are set inside the accelerator vacuum and signals are brought out to external electronics for processing. Using Eq. (4.28), for $r/a \ll 1$ and $\phi \sim 0$, the linear current density at a pickup with $y \sim 0$ is $j(x) \sim I_b/(2\pi a)(1 + 2x/a)$. The current on the pickup I_A due to a charge at small x is linear in x. The current I_B, located at $x \sim -a$ and $y \sim 0$, is compared and subtracted. For small x, $X_{AB} = (I_A - I_B)/[2(I_A + I_B)] \sim x/a$ and the comparison of the two BPM gives a measure of the x of the beam.

$$X_{AB} = (I_A - I_B)/2(I_A + I_B) \sim (x_b/a) \qquad (4.30)$$

The response of a BPM to a point current is explored in the App "BPM_XAB_YAC". A "Slider" is used to choose the location of the point, $d = r/a$ and ϕ. The BPM pickup subtends an azimuthal full angle α also chosen by "Slider". The current is found by integrating j, Eq. (4.28), over the pickup using the Matlab numerical utility "integral". The deviations of the pickup determination of x and y are displayed via numeric "EditFields". Specific results are shown in Figure 4.22. For $x/a < 0.2$ the agreement of x and X_{AB} is close.

Using the results above, a mapping of the errors in measuring x and y with four BPM pickups can be explored. This is done with the script "BPM_x_y_err" with plots of the x and y errors shown in Figure 4.23. With x and y BPM one can map the measured X_{AB} and Y_{AB} into the correct (x, y) of a beam in software by making a mapping. In any case, for a beam comfortably localized within a beam pipe and well centered, a measurement of the (x, y) location of a beam particle can be made. There is also a hardware solution which corrects the BPM by altering and extending the pickup geometry to linearize the signals, but it takes up additional space along the beam and will not be explored here.

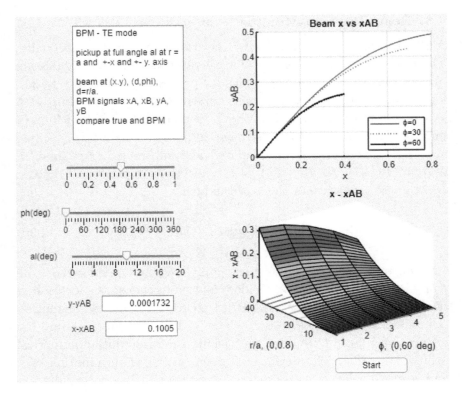

Figure 4.22: Plots for $x - X_{AB}$ as a function of x for a few ϕ values are made and the surface of $x - X_{AB}$ is shown for a range of (x, y) or $(r/a, \phi)$ values with $\alpha = 10°$.

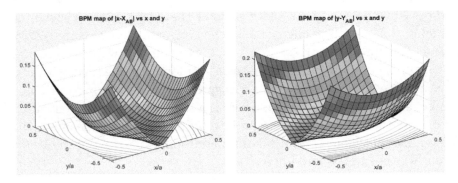

Figure 4.23: Plots of the absolute errors in x and y using a four pickup BPM configuration. The pickup full azimuthal angular coverage is fixed at $20°$. The error is smallest when the beam particle is near the origin at $(0,0)$.

4.9. Transverse Beam Position — ε, δ

The results on BPM can be extended from a single particle to the full transverse beam size. The measured shape is accurate for small excursions from the (0,0) of the beam pipe. For $(X_A-X_B)/(X_A+X_B)$ the largest error terms are of size $\sim(\sigma/a)^2$. There are also effects of a BPM with a finite frequency response or to a NR beam as shown later in the text. In addition there are the effects of beam dispersion. As mentioned already in Section 4.3, the measured beam r.m.s. is due to both transverse emittance ε and to momentum spread $\delta = dp/p$, whose effects depend on the location in the lattice specified by $\beta(s)$ and $\eta(s)$, $\sigma_x^2 = \varepsilon_x\beta_x(s) + [\eta(s)\delta]^2$ but typically are largest in the F quadrupole of the FODO.

Instead of a single charge, the App "Beam_Size_ x_y" generates 200 beam particles populated as Gaussians in x, y with mean x_o and y_o. The means and standard deviations are chosen by "Slider". The units for the Gaussians are for $a = 1$. The beam distribution and the BPM reconstruction are plotted in Figure 4.24 and can be compared. Indeed, the full beam is well reconstructed for small x, y excursions but should be mapped if the beam populates a substantial fraction of the beam pipe transverse area.

If a beam is not UR (linac or low energy cyclotron for example), then the image charge is not a delta function in time $\delta(t)$, nor is it fully transverse (TE) as seen in Section 4.5. The induced charge depends on the frequency response of the BPM due to the finite time of traversal of the BPM. A typical time interval is $(a/\gamma\beta c)\sim 1/\omega$ as shown previously in Section 4.5 and shown again in Eq. (4.31). There is no closed form solution as there was for the TE case, Eq. (4.28). However, a series solution exists as an expansion in terms of modified Bessel functions, $I_v(x)$. As is the case with most special functions, these too are available in Matlab as "besseli" and ten terms of the series are evaluated in the App.

$$(\omega a/\beta c\gamma) \qquad\qquad (4.31)$$

These solutions are displayed in the App "BPM_Pickup_notTE" and are shown in Figure 4.25. A single particle on the x axis, $d = x/a$,

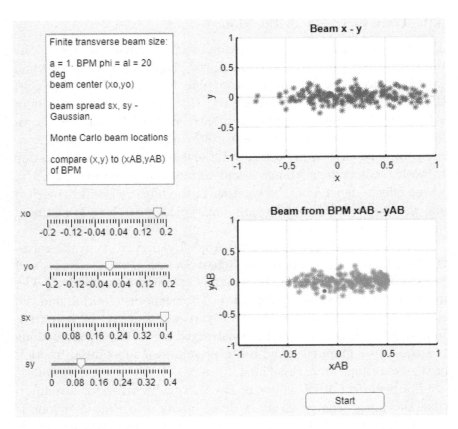

Figure 4.24: Comparison of a simulated transverse beam to a BPM reconstructed beam. The user can vary the properties of the beam, mean and standard deviation, in order to see how well the BPM reconstruct the centroid and r.m.s. values.

is chosen by "Slider" and tracked. The plot of the pickup current I_A and the difference of X_{AB} and x are shown as a function of ω in Figure 4.25. It is clear that signal processing of the pickup needs to pass frequencies somewhat below those indicated by Eq. (4.31) in order to get an accurate measurement of the beam position. A high pass filter may not be sufficient.

With a small BPM the bunch length and hence dp/p can be measured by using the time of passage of a bunch. In Section 4.4, the MR bunch length of 1.1 m for $dp/p = 0.1\%$ was only 3.7 ns.

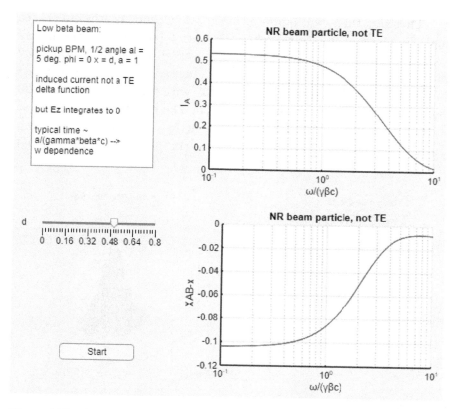

Figure 4.25:   Response of a BPM to a low velocity beam where the electric field at the BPM is not instantaneous nor is it fully transverse. Low frequency response needs to be preserved.

Accelerators may have even shorter bunch lengths, which means that an accurate direct measurement of the time of passage of a bunch is not always possible. However, using BPM at lattice positions with different dispersions allows one to determine the momentum spread in the beam which is proportional to the bunch length. For the Fermilab MR FODO the maximum $\eta(s)$ and $\beta(s)$ are at Q_F, while the minima are at Q_D. If the x standard deviation of a beam is measured at these two points, the emittance and the momentum spread of the beam can be determined assuming the lattice parameters $\eta(s)$ and $\beta(s)$ are known.

There are then two equations in the two unknowns, the emittance and the momentum spread, $\sigma_x^2 = \varepsilon_x \beta_x(s) + [\eta(s)\delta]^2$ and the utility "solve" is used. The FODO parameters were discussed in Sections 4.2 and 4.3 and displayed in Figure 4.8. Gaussian beams are generated in the App. "BPM_dpp_QF_QD_FODO" with typical results for 5000 beam particles are shown in Figure 4.26. The emittance was fixed at $3.8 \, \text{mm} * \text{mrad}$. The full 3×3 FODO matrices are used. The Matlab utility "histogram2" is used to produce the plots, sometimes called

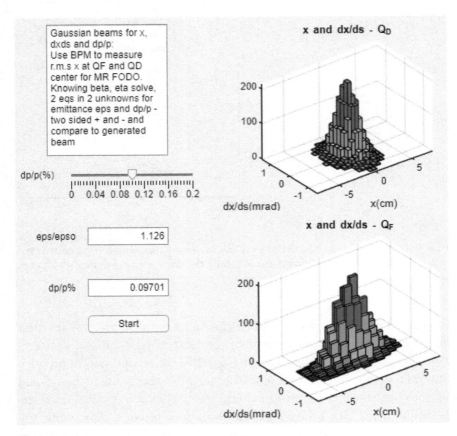

Figure 4.26:   FODO particles which are Monte Carlo generated as Gaussians in x, dx/ds, and dp/p. The r.m.s. in x for the center of Q_F and Q_D are used to determine ε and dp/p and compared to the values used in generation. The Matlab utility "std" is used to compute the r.m.s. values.

"LEGO plots" for obvious reasons. The solutions are accurate within the statistics, estimated to be $\sim 1/\sqrt{5000} = 1.4\%$ for small values of dp/p.

4.10. BPM and "Accelerator" Tune — μ, Q, β

Having looked at the induced charge on a BPM and a way to use two BPM to find the emittance and momentum acceptance, we turn to "accelerator" simulation. For our purposes an "accelerator" is simply 96 MR FODO with beams populated as Gaussians via Monte Carlo techniques. In this section, a single turn is considered and turn by turn sampling of the beam is assumed. The "accelerator" can have a variable number of sequential BPM spread around a fraction of the circumference. This deployment is not optimal. The location of the BPM will normally be where the size is a maximum, at the entrance to the FODO at the center of the Q_F quadrupole.

The tune is found in the App "Accel_Q_BPM2" with output shown in Figure 4.27 using the utility "fft" as shown with sunspot data previously, Figure 4.19. Only a single particle is tracked, which was generated with a Gaussian distribution in x and dx/ds. The number of contiguous samples is chosen by the user by writing into the "EditField". The calculated tune and the measured tune are shown in "EditFields"' For 96 BPM, one per FODO, the tune error is small $\sim 1\%$. For a smaller number of BPM the accuracy of the Q value deteriorates. The Q for which a particle is chosen to track, chosen using the "Start" button again to generate a new particle, does not change appreciably which shows that there is little dependence on the beam location.

There is a spread in the tune values of an accelerator because there is a finite momentum acceptance for particles in the beam. The different momenta have different quadrupole focal lengths, leading to different tune values. The chromaticity is defined to be the change in tune arising from a momentum shift, $\Delta p/p, Q' = \Delta Q/(\Delta p/p)$. The differential of a continuous quantity is indicated by d in this text while a shift in a value is indicated by Δ. Normally the quadrupole chromaticity is removed using sextupoles with the

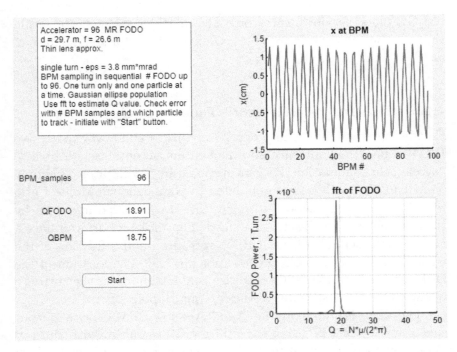

Figure 4.27:   Output of the App to make a simple accelerator model with BPM tracking of a single particle. The QFODO value follows from the MR lattice calculation and is compared to the BPM power maximum.

remaining chromaticity being due to the dipoles. The quadrupole effect is explored in the App "Accel_Q_Quad_Chromo2". In this case the number of BPM is chosen and again only one particle is picked from Gaussian distributions in x and dx/ds. The focal length is changed by a constant $\Delta p/p$, $f \rightarrow f(1 + \Delta p/p)$ with a specific result seen in Figure 4.28. The observed shift in Q for a 4% change in momentum is 4.4% $\Delta Q/Q$, 18.90 $\rightarrow$ 18.08, using the matrix element phase measurement and 4.0%, 18.75 $\rightarrow$ 18.0, for a one turn measurement of position with 96 BPM. With 96 BPM the tune shift is accurately measured in this simple model. The dipole matrix elements $M(1,3)$ and $M(2,3)$ are shown for reference but they have no effect on the betatron oscillation tune which is defined by the 2×2 transfer matrix relevant to the central momentum.

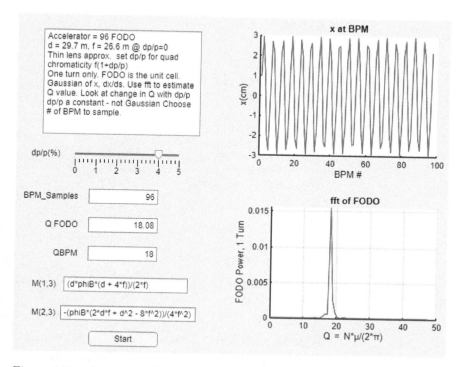

Figure 4.28: Output of the App "Accel_Q_ Quad_Chromo2" showing the effect of changing the quadrupole focal length on the FODO tune as calculated and measured in a series of BPM.

If the accelerator has sextupole corrections of the quadrupole chromaticity, the dp/p of the beam can be measured by observing the tune shift when the dipole field is changed while keeping the r.f. unchanged. The beam momentum is then changed while the orbit remains fixed so that $\Delta p/p = \Delta B/B$. Alternatively, the r.f. can be changed which changes the $\Delta p/p$ in proportion. The relationship between the particle momentum and the r.f. is found by measuring the tune shift as a function of the change in the r.f. The beam momentum shift can be measured, assuming that the slip factor S, defined in Eq. (4.20), is known.

$$\Delta p/p = (-\Delta\omega_{rf}/\omega_{rf})/S \qquad (4.32)$$

```
>> beta_Quad_Kick_M1T

M1T =

[   cos(u1),  bet*sin(u1)]
[  -sin(u1),       cos(u1)]


M1Tk =

[ cos(u1) + bet*k*sin(u1),  bet*sin(u1)]
[       k*cos(u1) - sin(u1),      cos(u1)]


u1k =|

acos(cos(u1) + (bet*k*sin(u1))/2)
```

Figure 4.29:   The FODO transfer matrix with and without the added "kick" and the phase change induced by the "kick".

One can also use BPM at Q_F and look at the one turn transfer matrix, M1T. Assume one can excite a single quadrupole at Q_F with a kick, $k = 1/f$, and then measure the resulting tune shift using the FFT as described above or by some other method. The tune change can be used to find the β value at the BPM. An example using the standard MR FODO is done using the script "beta_Quad_Kick_M1T" with output shown in Figure 4.29. The kick is added to the initial FODO element by multiplying by the matrix $[1\ 0\ ;\ k\ 1]$. Then using the one turn transfer matrix, Eq. (4.2), and finding the change in phase, the lattice beta function at Q_F is determined. The phase shift depends on β at Q_F. The result for different values of k is shown in Figure 4.30 where the slope of the tune shift change is β at $Q_F \sim 100$ m.

4.11. Quad Kick and Beta — β, σ

In Section 4.9, two BPM, one at Q_F and a second at Q_D, were used to extract the beam emittance and dp/p as in Figure 4.26. Then, in

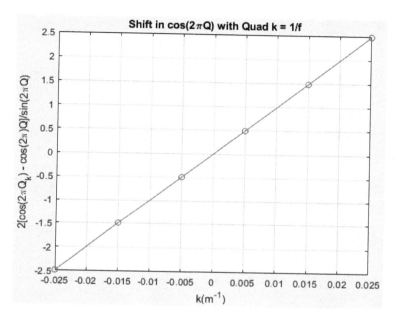

Figure 4.30:   Plot of the change in cosine of the phase at Q_F after one turn as a function of k. The slope is the value of β at Q_F or $\sim 100\,\text{m}$ for the MR case.

Section 4.10, a single BPM was used with FFT to extract the tune Q as seen in Figure 4.27. The tune shift due to a shift in quadrupole focal length was displayed in Figure 4.28. Then a single BPM with a kick at a single quadrupole was used to extract the value of β at that point using the induced tune shift in Figures 4.49 and 4.30. Now more complex situations with multiple BPM are explored to extract other or complementary information about the properties of the beam. A schematic of the procedure is shown in Figure 4.31 where a kick is applied to the beam and the resultant disturbance is recorded elsewhere in the accelerator lattice.

For a single BPM as shown in Figure 4.31, the response to multiple values of the quadrupole kick, k, can be used. With a fit to a quadratic form in k, one can extract, using the beam matrix at the single BPM, properties of the beam. For example, assuming the lattice parameters are known, the emittance of the beam can be

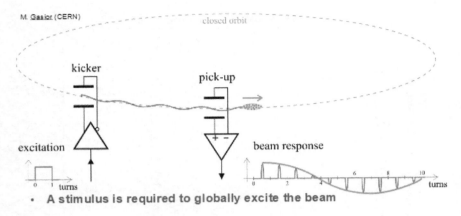

Figure 4.31: Schematic of the result of a kicker at a single location in the lattice and the response recorded at a second location.

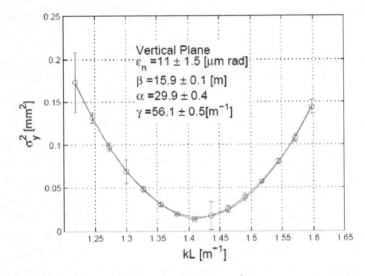

Figure 4.32: Fit to data of the beam matrix $\langle y^2 \rangle$ at a BPM as a function of $k = 1/f$.

measured using the values of the fit parameters. This procedure with real data is shown in Figure 4.32.

The beam matrix, σ, is $\varepsilon\Sigma$ as shown in Eq. (4.6), while the transformation to another location is governed by the transfer matrix

```
>> Bump_Q_BPM_Beam_Size
Transfer Matrix M from Q to BPM
Initial Beam Matrix S at Q
Beam 11 = <x^2> at BPM
M12 (S12 (M11 + M12 k) + M12 S22) + (M11 + M12 k) (S11 (M11 + M12 k) + M12 S12)

k^2 Term = S11(M12^2) ~ <x^2> at Q
k Term = 2[M12][S12M12+S11M11] ~ <xdx/ds> at Q
Constant Term, Compare to k = 0, ~ <(dx/ds)^2> at Q
det(S) = Emittance at Q
```

Figure 4.33: Output of the symbolic calculation of the 11 components of the beam matrix $\langle x^2 \rangle$ at the BPM after a bump quadrupole.

M, as $\Sigma_f = M\Sigma_i M^T$, Eq. (4.7). This transformation is done symbolically and explicitly using the script "Bump_Q_BPM_Beam_Size" with output shown in Figure 4.33. There are terms in the beam matrix at the BPM both quadratic and linear in k as well as a constant term which corresponds to the beam without the bump perturbation. The quadratic term is proportional to $S_{11} = \langle x^2 \rangle$ which is measured at the BPM and the transfer matrix is assumed to be known having performed the lattice calculations. The three parameters in the fit can be used to find ε, β, and α as shown in Figure 4.32.

The use of the quadrupole "kick" technique with the resultant beam size sampled in a single BPM downstream of the "kick" is explored in the App "Accel_Quad_Kick" with output shown in Figure 4.34. In thin lens approximation the "kick" occurs at the start of a MR FODO and the results are sampled at Q_D, a "waist" with mimimal beam size in order to be most sensitive to the "kick" effects. The number of beam particles used is chosen by "Slider" as is the size of the "kick". The sigma of the beam at the waist is shown using "EditFields" using the Matlab utilities "mean" and "std" to find the mean and standard deviation. The initial beam is a Gaussian in x and dx/ds. In the plots, a "movie" of the first few beam particles shows the position through the three downstream FODOs. The second plot is the location of beam particles at the "kick" located at the start of the first FODO.

Some results from the App shown in Figure 4.34 are collected and displayed in a plot in Figure 4.35 using the script

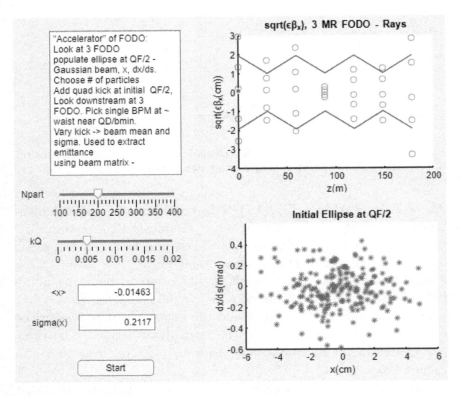

Figure 4.34:   Result of a particular quadrupole "kick" at a MR FODO at the Q_F location. The beam r.m.s. at a downstream Q_D is shown for a value of the "kick" k_Q chosen by the user selected "Slider" value.

"Quad_Kick_QF_QD_sigmax. The execution of the previous app is a bit slow for a large number of particles, so the results are supplied in the script. A parabola, shown in red, has been added to the points. It is in no sense a fit to the expression shown in Figure 4.33 and is only made to be indicative of a quadratic term in the beam r.m.s. at Q_D. The shape is similar to that of the data displayed in Figure 4.32.

Another impulse to apply to the beam is to change a single dipole field in analogy to a kick to a quadrupole. Assume this is done near to a BPM. The angular kick from the dipole, with change in angle θ_B, creates a new closed orbit. The one turn

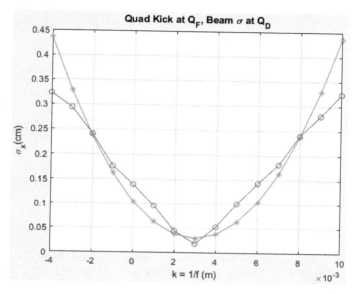

Figure 4.35: Plot of tabulated results of the App "Accel_Quad_Kick" for the r.m.s. of the beam at Q_D as a function of the strength of the "kick", k, located at a Q_F 1 and 1/2 FODO upstream. The parabola in red is merely indicative. No statistical errors arising from only 200 beam particles have been assigned or plotted.

transfer matrix was I by definition, the identity matrix, since the accelerator is cyclic, but has been changed to M. One can then solve for the closed orbit change, for x_B and $dx/ds = x'_B$ at the dipole. The results are shown in Eq. (4.33). They are derived using Matlab utilities. The one turn transfer matrix was defined already in Eq. (4.2) with $\mu = 2\pi Q$. The algebra is done using the script "Dipole_Kick_BPM_beta". The Matlab utility "eye" populates the identity matrix. The utility "solve" solves the two equations for x_B and x'_B. Finally, "simplify" and "subs" are used to clean up the symbolic solutions. Some of the steps are displayed in Figure 4.36. Using a BPM or other instrumentation to find the new closed orbit, the lattice function β at the dipole can be determined. Assuming the Q is measured, and the dipole kick is also known, the closed orbit shift is directly proportional to β. The amplification by β is large. A dipole kick of only 0.2 mrad causes a closed orbit shift with a scale

```
>> Dipole_Kick_BPM_beta

M1T =

[       cos(u) + a*sin(u),              b*sin(u)]
[ -(sin(u)*(a^2 + 1))/b, cos(u)  - a*sin(u)]


equs =

                x*(cos(u) + a*sin(u) - 1) + b*dxds*sin(u)
    thB - dxds*(a*sin(u) - cos(u) + 1) - (x*sin(u)*(a^2 + 1))/b


xB =|

(b*thB)/(2*tan(u/2))


dxdsB =

(thB*(sin(u/2) - a*cos(u/2)))/(2*sin(u/2))
```

Figure 4.36: Output of the steps taken to solve for the shift in the one turn transfer matrix due to a dipole angular kick and the subsequent closed orbit changes, x_B and dx_B/ds.

of $\beta\theta_B/2 = 1\,\text{cm}$ at the Q_F location.

$$(M - I) \begin{bmatrix} x_B \\ x'_B \end{bmatrix} = \begin{bmatrix} 0 \\ -\theta_B \end{bmatrix}$$

$$x_B = (\theta_B/2)\beta/\tan(\mu/2)$$

$$x'_B = (\theta_B/2)[1 - \alpha/\tan(\mu/2)]$$

(4.33)

4.12. Multiturn and Phase — Q, β, α

In a situation where a single BPM takes data for many turns of a beam, there is information on the tune that can be determined. Previously, only a single turn with many BPM was considered in finding Q. A multiturn model is created in the App "Accel_Multi_Turn". For the MR FODO the phase advance is $\mu = 1.2376\,\text{rad} = 70.707°$. For one turn or 96 FODO, $\mu_1 = 0.5662$ since $96\mu = 118.8096$ or $Q = 18.909$. Using Matlab "acos" for the one turn matrix trace, the Q_1 angle is 0.5662 rad or 32.44°. This wrap around phase increments each turn until $360/32.44 \sim 11.097$ turns as seen in the plot shown in Figure 4.37 for m turns of the cumulative phase at a single BPM. This

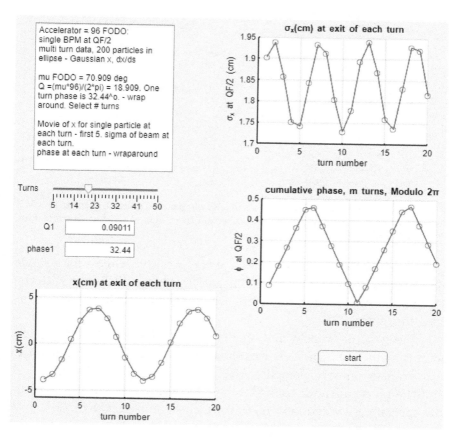

Figure 4.37: Multiturn behavior of the MR "accelerator". Individual particles are displayed as well as the r.m.s. of the "beam" of 200 particles and the cumulative phase as calculated in Matlab.

behavior is seen in the "movie" of the first five individual particles (lower left plot) and a small residual effect is also evident in the plot of the r.m.s. of the 200 particles in the beam.

An FFT analysis with 200 particles in a bunch, tracked over 100 MR turns, is available in the script, "FFT_Multiturn". A single particle can be used, or the r.m.s. of the bunch. Results for both a single particle and the bunch r.m.s. are shown in Figure 4.38. For the single particle the power maximum is at 11.1111 turns, while for the r.m.s. the power is a maximum at 5.5556 turns which is half the single

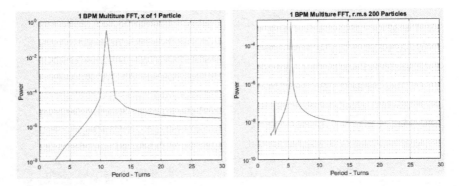

Figure 4.38:   Results of FFT analyses of a single particle going 100 turns in a MR "accelerator", left, or a bunch characterized by the r.m.s. of the bunch containing 200 particles.

particle result because the r.m.s. is always positive while x for a single particle can be both positive and negative. The results agree with the calculation of the wraparound phase at $11.097 = 360°/32.44°$ turns seen in Figure 4.37.

The general transformation defined by the transfer matrix M with the final value of β, β_f was shown in Eq. (4.7) and in Figure 4.4. The transfer matrix elements M_{11} and M_{12} as a function of β_f, β and the phase advance from β to β_f, $\Delta\psi$ were shown in Eq. (4.8) in the special case where $\alpha = \alpha_f = 0$ which obtains at Q_F or Q_D.

$$\beta_f = \beta M_{11}^2 - 2\alpha M_{11} M_{12} + \gamma M_{12}^2$$

$$M_{11} = \sqrt{\beta_f/\beta}(\cos \Delta\psi + \alpha \sin \Delta\psi)$$

$$M_{12} = \sqrt{\beta_f \beta} \sin \Delta\psi$$

$$\tan \Delta\psi = M_{12}/(M_{11}\beta - M_{12}\alpha)$$

(4.34)

Using three BPM the phase at each can be determined by simple lattice calculation for β, α, M_{11} and M_{12}. For the relative phase from BPM_1 to BPM_2 and BPM_1 to BPM_3, the transfer matrices M and N are used which are known from the lattice design. That then allows one to determine β and α at the first BPM. The calculations are made symbolically in the App "Multiturn_beta_alpha_3BPM". There are two equations with two unknowns for the lattice parameters

α and β at the first BPM in terms of the two transfer matrices and the two phase advances. They are solved symbolically using "solve", and displayed both symbolically and numerically. A specific layout is chosen with BPM_1 at $Q_F/2$, BPM_2 at Q_D and BPM_3 at either the next $Q_F/2$ or the next Q_D. Since $\alpha = 0$ at those locations the algebra is simplified. The choice for the third BPM location is made using the "DropDown" utility. The FODO parameters can also be varied using the "Sliders" in order to change the lattice phases. Specific results are shown in Figure 4.39.

Besides using BPM to explore the lattice characteristics of an accelerator, the BPM is an extremely useful diagnostic tool. Modern accelerators have very extensive diagnostics made available by deploying a large suite of accelerator instrumentation. There are

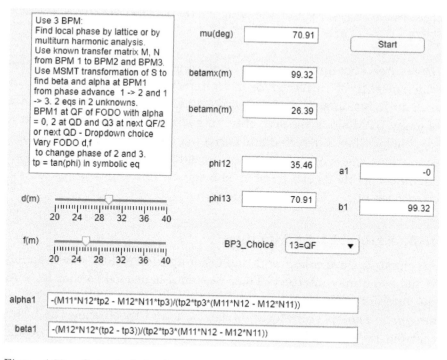

Figure 4.39: Output of the App used to find α and β at the start of a FODO in terms of the phases and transfer matrices to the two downstream FODO elements.

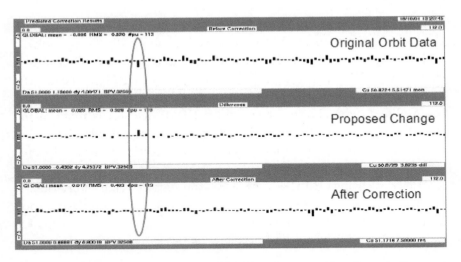

Figure 4.40:   Illustration of the use of BPM to detect and localize misalignment of accelerator elements. Subsequent verification of the alignment correction is also shown.

accelerator startup issues such as misaligned accelerator elements, which need to be corrected for in an iterative fashion. An example of such an iteration appears in Figure 4.40 which shows the disposition of many BPM along the accelerator structure. A beam deviation at a specific location is detected and corrected for. After a number of such corrections the accelerator can be considered to be aligned. Optical alignment methods can be used as a source of initial settings or as a crosscheck of the BPM procedures.

4.13. External Loss Monitors

Accelerators have collimators and other apertures that some fraction of the beam may intersect. Their existence is needed so that a beam can interact with these limiting apertures in a controlled fashion as designed. The aperture location can be found by steering the beam centroid by a controlled amount and seeing when large losses occur. A large "spray" of secondary particles, detected by devices external to the beam pipe signals that some part of the beam has intersected a limiting aperture and caused a shower of secondary particles, some of which are detected in the loss monitors. Typically beam line techniques can be applied to accelerator exterior measurements of losses of the accelerator beam at limiting apertures or collimators.

The simplest and most robust loss monitor is an ionization chamber. It is a sealed gas volume with an applied electric field. All the ionization of the gas is collected and detected as a signal current. There is no gain which makes this type of detector very rugged. Other loss monitors could be a PWC or a scintillation counter. These detectors were defined previously. However, the radiation field even external to the accelerator beam pipe may be so intense that these devices cannot operate properly.

An amusing example of an external set of loss monitors is the use of an entire detector at the LHC, in this example the CMS and ATLAS detectors. The appearance of a massive shower, seen in Figure 4.41, in the calorimeters and muon chambers of both experiments is a strong indication that the LHC beam(s) have scraped a limiting aperture. The detectors for colliding beams at the LHC were then also able to contribute to the LHC machine diagnostics as "loss monitors".

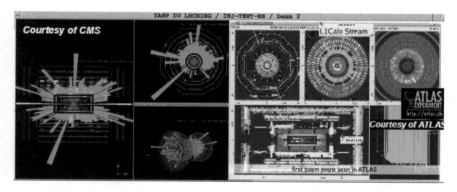

Figure 4.41:   Detector displays of the ATLAS and CMS calorimeters and muon detectors during a time when the LHC beam(s) were scraping an upstream limiting aperture. The more delicate tracking systems were turned off as a safety precaution.

4.14. Beam Current Transformer

So far, the induced charge on an electrode within the beam pipe vacuum has been the instrument for beam detection, the BPM. However, since the BPM is inside the accelerator vacuum it means a problem with it may necessitate a long access and maintenance period. It is desirable to have a device outside the beam pipe, which does not interfere with the beam itself yet makes robust and reliable measurements of the properties of the beam. A common solution to this problem is the beam current transformer (BCT). In the BCT a beam current I_b, creates azimuthal magnetic field, $B_b \sim (\mu_o I_b)/(2\pi r)$ which the transformer, an external toroid, converts to signal current, I_s. The beam itself is the primary of the transformer and has only one turn. The toroid with N_s turns is the secondary. A load resistor R_s develops the secondary voltage, $I_s = I_b/N_s$ and the voltage is $V_s(\omega) \sim I_b(\omega)R_s/N_s$ if the impedance is ∼real. A schematic of the BCT is shown in Figure 4.42. Note, however, that there is, of necessity, a break in the wall metal and an r.f. bypass path around the BCT ceramic pipe. That bypass causes a change in the smooth impedance of the continuous metallic beam pipe.

There is no d.c. coupling allowing a constant current flow so that the signal voltage is frequency dependent with the low frequencies lost. At typical frequencies for ω of 1–100 MHz, the impedance is approximately a real constant, $Z = V_s/I_s = R_s/N_s$. In Figure 4.43, the interior of a BCT is shown where the break in the beam pipe metal is visible as are the ferrite toroids surrounding the beam pipe bypass. Figure 4.44 is a photo of an assembled set of BCT toroids which indicates the typical scale of the device.

The BCT frequency response needs to be considered further. The toroid has an inductance which depends on the ferrite permeability, μ, the number of secondary windings, N_s, and the ratio of the inner and outer toroid radii, a and b. Considering only the secondary inductance L_s and resistance R_s, the secondary current appears in Eq. (4.35). There is a characteristic L–R time constant, τ_L. As mentioned previously j is $\sqrt{-1}$. The inductance imposes a low

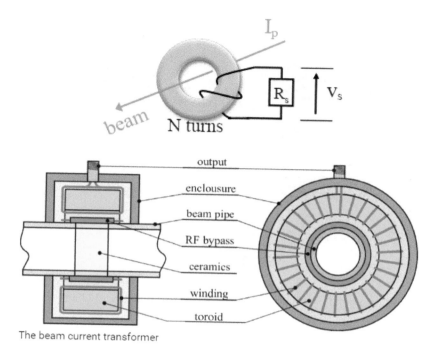

The beam current transformer

Figure 4.42:   Schematic of a BCT showing the ceramic break in the beam pipe
and the r.f. bypass. The secondary toroid is also shown as is the enclosure used
to shield the BCT.

Figure 4.43:   Photo of an actual BCT showing the break in the beam pipe and
the outer ferrite toroids.

Figure 4.44:   Photo of an assembly of BCT toroid windings.

frequency cutoff at $\omega \sim 1/\tau_L$.

$$L_s = \mu N_s^2 \ln(b/a)$$
$$I_s(\omega)/I_b \sim j(\omega\tau_L)/N_s[1 + j(\omega\tau_L)] \qquad (4.35)$$
$$\tau_L = L_s/R_s$$

With the additional consideration of stray capacity and load resistor R_s, the signal sees an R–L–C circuit. The midrange for a real impedance, $Z \sim R_s$, is $\sim 1 - 100$ MHz for typical values of the parameters such as $\mu > 10^4 \mu_o$. At high frequencies, $\omega\tau_L \gg 1$, the secondary current approaches I_b/N_s. With a winding capacity and/or signal cable capacity the secondary capacitance must also be considered at low frequencies. There is an additional term in the impedance, now with both a fast time constant, $\tau_C = R_s C_s$, and a slow constant τ_L. In terms of time and not frequency the slow falloff goes as $V(t) \sim [I_b(t)/N_s]R_s \exp(-t/\tau_L)$. The secondary capacity imposes a high frequency cutoff. The full impedance appears in Eq. (4.36). Another schematic of a BCT appears in Figure 4.45

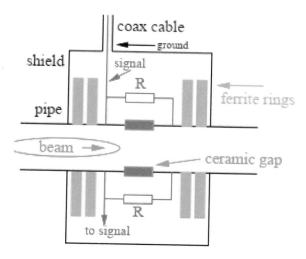

Figure 4.45:   Schematic of a BCT with L_s and R_s coupled to a coaxial cable to route the signal to external instrumentation. The stray ferrite and cable capacity, C_s, is not indicated.

which emphasizes the signal shielding and the coupling of the signal to a shielded coaxial cable, as already discussed in Section 3.18.

$$\tau_L = L_s/R_s, \tau_C = R_s C_s$$
$$Z(\omega)/R_s = j(\omega\tau_L)/[1 + j(\omega\tau_L) - \omega^2\tau_L\tau_C] \qquad (4.36)$$

A view of the BCT impedance is produced in the App "Z_RLC_App". The user chooses the ferrite inductance and the stray capacitance using the "Sliders" and a 50 Ω load resistor is assumed to match to a standard coaxial cable. With a ferrite permeability factor of $\sim 10^4$ and a stray capacity of a few pF a reasonable range of frequencies where the impedance is real and $\sim R_s$ is obtained, up to 10 GHz as shown in Figure 4.46. To set a scale, a single particle with a γ factor of 400 has an electric field active time, at a pickup radius of 5 cm, about 0.42 ps, or with frequencies of about 24,000 GHz. Indeed, individual particles are not easily resolved in time, but the aggregate bunch length is set by the r.f. system and can be resolved using different techniques, as will be discussed later.

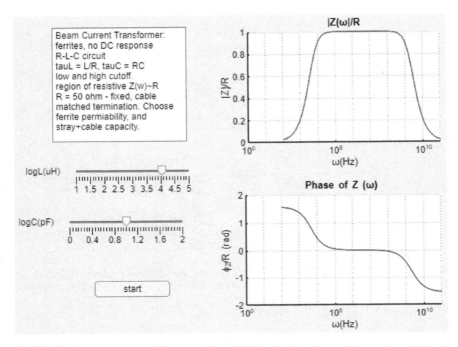

Figure 4.46: Plots of the magnitude and phase of the impedance of an $R-L-C$ circuit, such as a BCT when the inductance is large, and the capacitance is small. There is a substantial range of frequencies where the impedance is essentially real and equal to R.

4.15. Beam Emittance — ε

The full phase space of the beam is described by the transverse position and velocity in x and y, $(x, dx/ds = x')$, (y, y') and the bunch length and momentum acceptance, (z and dp/p). At this level of approximation the transverse betatron oscillations in x and y are independent with no coupling between x and y. The chromaticity, with finite dp/p, and bunch length in z is determined by the r.f. of the accelerator. The beam matrix, $\sigma = \varepsilon\Sigma$, defined in Eq. (4.6), depends on the location in the accelerator and has a finite area, $\pi\varepsilon$, defined by the emittance ε. The ellipse area in (x, x') and (y, y') is preserved if the beam energy is constant. The ellipse boundaries are shown once again in Figure 4.47. They are the r.m.s. of x and x' for the beam matrix σ and the ellipse is not a sharp boundary.

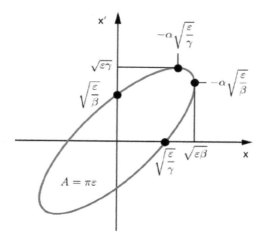

Figure 4.47:   Beam ellipse in terms of the lattice parameters α, β, and γ. The ellipse boundary indicates the r.m.s. of the beam. For example, $\langle x^2 \rangle = \varepsilon \Sigma_{11}$.

Rather it indicates the region where 68%, one "sigma", of the beam particles are contained. The transformation matrices, M, have unit determinant, $|M| = 1$ and the inverse is the transpose, $M^{-1} = M^T$ as is common for transformations in conservative systems in classical mechanics. Because of that property, $|\sigma| = \varepsilon$ and $|\Sigma| = 1$.

In previous sections, the use of BPM to determine properties of the beam has been explored by looking at a few examples. In Section 4.9, the use of two BPM allowed the determination of ε and $\delta = dp/p$. In Section 4.10 many BPM were used to find the tune, Q and a quadrupole kick induced a tune shift that allowed the extraction of $\beta(s)$ at a BPM. In Section 4.11, a "kick" induced a quadratic increase in the beam size downstream. In Section 4.12, a single BPM sampling many turns also determined the accelerator tune, Q. Also in that section, two BPM downstream of a first BPM and known phase advances were used to extract β and α at the first BPM. In this section, a few more ways to use BPM to interrogate the beam are discussed.

The ellipse shape, defined by α, β, and ε, is calculable for a specific accelerator design. The overall beam size is determined experimentally using a quadrupole to "kick" the beam and then to observe the effects down-stream of the quadrupole at a single BPM. The results were shown already in Figure 4.32 quoting actual data and in Figure 4.35 using a simple Monte Carlo 96 FODO MR model. Dipoles have a bend angle which depends on momentum. Off-momentum particles are not on the accelerator reference design orbit. That means their ellipse for off-momentum particles is not aligned to (0,0) in the bend plane; the ellipse is off center for off-momentum particles. This increase in beam size gives a measure of the dispersion/chromaticity of the accelerator as was shown in Figure 4.26. A schematic of the technique is shown in Figure 4.48 where some adjustable elements, quadrupole or dipole for example, modify the unperturbed beam, and those changes are sampled at a single BPM. The beam FWHM should have the dp/p momentum spread removed, where FWHM $= 2.36\sigma$ for a Gaussian beam with r.m.s. of σ, in order to extract the lattice parameter, such as $\beta(s)$. In that case, $\varepsilon_x \beta_x(s) = \sigma_x(s)^2 - [\eta(s)\delta]^2$.

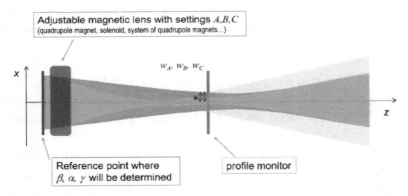

Figure 4.48:   Schematic of a general beam excitation using a single element near where lattice parameters are to be determined. The results are sampled at three values of excitation at a single BPM. In the specific case of a quadrupole it has over focused the beam from its unperturbed value as shown in blue.

An alternative technique is to use three BMP near a waist and use the three measurements to find the beam parameters, ε, β, and α, or the beam matrix σ. A schematic of the procedure appears in Figure 4.49. The beam is sampled in three BPM downstream of a specific accelerator element. The three measurements, with

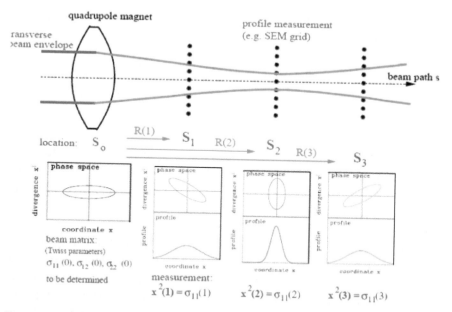

Figure 4.49:   Schematic of the use of three BPM downstream of an accelerator element in order to find the beam matrix at that element. The transfer matrices are labeled R in this figure whereas they are consistently labeled M in the text. The ellipse has the parameter α, the "tilt", $= 0$ at s_o and s_2.

knowledge of the transfer matrices to the three BPM, are sufficient to determine three quantities, for example the full beam matrix. In this specific case the element is a Q_F, so the beam size is largest there. Traveling to the BPMs, the beam size shrinks to a "waist" and then begins to diverge. In this case there are no intervening accelerator elements, so the transfer matrices are particularly simple, $M = [1\ 0\ ;\ d\ 1]$ where d is the distance to the particular BPM.

In general, the transformation of the σ_{11} component of the beam matrix from location o to location 1 has been shown previously and appears now in Eq. (4.37). The simplest case is to use only drift spaces, all of length d, between the BPM. Set BPM at $z = d$, $2d$ and $3d$. The BPM r.m.s. values and the lattice d give three equations in three unknowns which are the elements of the initial beam matrix. The emittance is determined by these three elements. For a simple drift space of length d, the last expression in Eq. (4.37) can be used.

In general one needs the calculated transfer matrix elements $M(1,1)$ and $M(1,2)$.

$$\sigma_{11}^1 = M_{11}^2 \sigma_{11}^0 + 2M_{11} M_{12} \sigma_{12}^0 + M_{12}^2 \sigma_{22}^0$$

$$\varepsilon = \sqrt{|\sigma|}, \sqrt{\sigma_{11}} = \sqrt{\langle x^2 \rangle} = \varepsilon \beta \qquad (4.37)$$

$$\sigma_{11}^1 = \varepsilon \beta_1 = \sigma_{11}^0 + 2d\sigma_{12}^0 + d^2 \sigma_{22}^0$$

The solution in a more general case is illustrated using the App "Three_BPM_Waist". The standard MR FODO is used. The beam at the start of the FODO is populated as a Gaussian as has been done before using Monte Carlo methods with $\alpha = 0$. Three downstream locations in the FODO were chosen, the first drift, the Q_D, the "waist", and the end of the second drift. The BPM r.m.s. at the three locations is evaluated and the three linear equations are solved for using the Matlab numerical utility "linsolve" which is used for a system of linear equations. The transfer matrices to the three BPM are assumed to be known and the three r.m.s. x distributions at the three BPM are calculated as input. The three elements of the beam matrix are calculated and the beam emittance is found assuming that the value of β at the start of the FODO is known. The emittance can be varied by use of a "Slider" and compared to the solved for value shown in an "EditField". A specific result is shown in Figure 4.50. The ellipses at the three BPM are plotted and the three r.m.s. x values are displayed numerically. They are analogous to the schematic plots shown in Figure 4.49. The solved for three parameters at $Q_F/2$ are also shown in units of centimeter and mrad.

4.16. Luminosity Measurement

An accelerator can also be used, after achieving its peak energy, as a storage ring or as a collider. A single ring can house counter-rotating beams of particles and anti-particles such as protons and anti-protons, as at the Fermilab Tevatron, or electrons and positrons, as at the LEP collider at CERN. With two rings, protons can be made to collide with protons, as at the LHC at CERN. In a storage ring the quantity used to find the cross sections for a process by

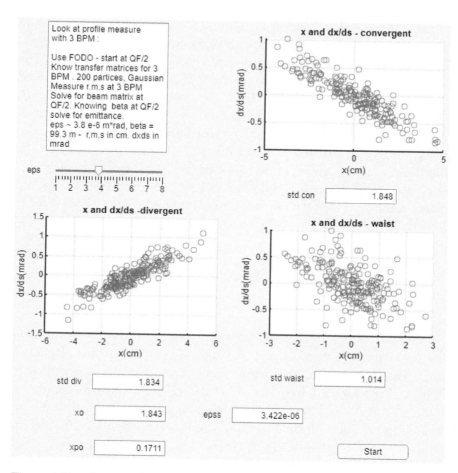

Figure 4.50:   Output of the results of using three BPM to determine the beam matrix at an initial location. The limited number of beam particles leads to a statistical spread in the values of the found initial beam matrix elements.

measuring the reaction rate is the luminosity. It is analogous to the factor $(N_A \rho L/A)$ which connects a cross-section to a mean free path if a target is at rest in the laboratory frame and indicates the number of target nuclei per unit transverse area. The rate, R, then follows from measuring the incoming flux of particles and is $R = dN/dt = \text{flux}$ times cross-section. The situation differs in a collider which is in the center of momentum (CM) frame.

In the case of a collider, for Gaussian beams with r.m.s. of σ in y (and x) of equal value for the two beams, the peak luminosity, L_o, is found when the beams overlap as completely as possible. The luminosity must be measured to extract the cross section, called "σ" here to avoid confusion, from the reaction rate, $R = dN/dt$. For beams labeled 1 and 2, with numbers of particles in a bunch N_1 and N_2, rotation frequency f, and number of bunches N_b the peak luminosity possible is shown in Eq. (4.38). To achieve a high luminosity the beams must be squeezed, $\sigma^2 \sim \varepsilon\beta$, and the value of β must be made as small as possible. This is accomplished by special insertions into the FODO structure of the accelerator, called "low β insertions".

$$dN/dt = L''\sigma''$$

$$L_o = (N_1 N_2 f N_b)/4\pi\sigma_x\sigma_y \tag{4.38}$$

$$L(\Delta y) = L_o e^{-\Delta y^2/2\sigma_{\Delta y}^2}, \sigma_{\Delta y} = \sqrt{2\sigma_y^2}$$

Numerically, for the LHC the vertical and horizontal beam sizes are $\sigma \sim 17\mu$m. The number in a bunch is $\sim 1.1 \times 10^{11}$ protons. Bunches are spaced by 25 ns, with some gaps. The number of bunches is 2808 and the rotation frequency is 11.245 kHz. The design luminosity is then 1.05×10^{34} (s/cm^2). The total number in a bunch can be found using a set of BCT as discussed previously. The frequency is set by the accelerator circumference. In principle the beam sizes could be measured using BPM and in fact BPM provide very useful checks.

However, because of the small beam sizes, the most accurate measurements of luminosity come from a technique called "Van der Meer (VDM)" scans" after the physicist who proposed them 50 years ago. Indeed, Simon Van der Meer received a Nobel Prize for Physics in 1984 for his many accelerator innovations. The idea is to use bump magnets (dipoles) to displace the 2 beams vertically by equal and opposite amounts for a total displacement of Δy. The spatial distribution of the luminous region does not move in that case so that detection efficiencies for a normalizing process do not change. The only thing that changes, in principle, is the actual luminosity or the reaction rate, as shown in Eq. (4.38). By measuring the rate of a process as a function of the y displacement, the beam size in y (and x) can be inferred and the luminosity measured.

The LHC bump distances at the interaction points are shown in Figure 4.51. The Gaussian rate decrease with the bump displacements is evident. The equal and opposite vertical displacements are tracked by BPM and are also shown in Figure 4.51 as a function of the bump magnet settings — calibrated in millimeter at the luminous region. The luminosity at the LHC is measured to an accuracy of about 2% at present using BCT to find the number of particles in the bunches and VDM scans to find the r.m.s. beam sizes. The error is dominated by systematic errors in the BCT. Direct measurement of the beam sizes using silicon detectors is also used as a cross-check of the VDM data since silicon detectors can have spatial resolutions comparable to or less than the transverse beam size as discussed previously.

The p–p elastic scattering process, Rutherford scattering, is a purely calculable process so that measuring the reaction rate as a function of the momentum transfer, $p\theta$, measures the luminosity directly. Treating the proton as a point particle with no structure at these very low values of the momentum transfer, the differential cross section, Eq. (3.10), appears in Eq. (4.39). In obtaining Eq. (4.39) a SR relationship for the NR calculation of Eq. (3.10) must be made since $T = p^2/2m$ is not a SR quantity. The replacement $T \to p\beta c/2$ was already mentioned in the discussion of ionization energy loss. The rate as a function of scattering angle, Eq. (4.39), depends only on known constants, the beam momentum and the luminosity. The α^2 factor from the two proton vertices involved the virtual photon exchange is expected.

$$dN_{el}/d(p\theta)^2 = L_o(4\pi\alpha^2)(\hbar^2/\beta^2)/(p\theta)^4 \qquad (4.39)$$

Detectors can also observe not only the p–p collisions which are very localized to the small interaction point, but also the p collisions of each beam with the residual or injected gas in the vacuum pipe. Assuming the residual gas density is known, in this case by injecting a known amount of neon gas, the size and N of the individual beams

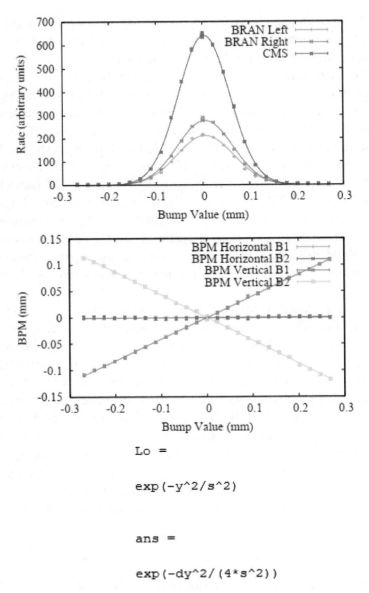

Figure 4.51:   Rate for a process at the LHC as a function of beam displacement. The equal and opposite vertical beam displacements are tracked and confirmed using BPM. There is also the output of a Matlab snippet for the product of two Gaussians, each with r.m.s. of s, displaced by $dy/2$ and $-dy/2$ where L_o is for $dy = 0$ and "ans" is $L(dy)/L_o$, Eq. (4.38).

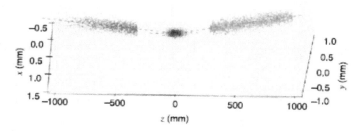

Figure 4.52: Collisions of the two proton beams at the LHC with injected neon gas are shown in red and blue. The beam–beam collisions are shown in black.

can be reconstructed. A picture of events identified with each beam individually is shown with red and blue points and the beam–beam collisions appear as black points in Figure 4.52.

Other processes than elastic scattering may not be precisely calculable from first principles, but their relative rates can be used to track luminosity variations during different accelerator operating conditions. Also used are wire scanners which intercept the beam causing a spray of secondary particles. These are measured by detectors both inside and outside the beam pipe. Alternatively, a screen with a set of small holes can be used to intercept the beam in both x and y, and the detected beam can be used to ascertain the transverse beam profile. Another direct method is to insert a screen of scintillating material directly into the beam and record the scintillation light, with a TV camera, for example. Secondary emission grids can be placed inside the beam pipe and the emitted current can be read out to reconstruct the beam size. The methods which utilize the insertion of a device into the beam are all at least partially "destructive" in that the beam itself is usually somewhat disrupted.

The strong interplay of accelerator and experimental techniques is very notable. It is for this reason that treating detector, beam and accelerator instrumentation seamlessly has been attempted in this text. The techniques apply the same physical principles and the three disciplines are synergistic.

4.17. Longitudinal Acceptance — δ, η

The longitudinal acceptance of the accelerator was derived in Eq. (4.25), $\delta^2 = (dp/p)^2 \sim (2qeV_o)/(\pi hScp_o)$ for zero synchronous phase angle — coasting beam. The finite acceptance in momentum arises from the limited region of phase stability within the r.f. cycle. A schematic example is shown in Figure 4.53. For the MR example, the dp/p was estimated at 150 GeV. The synchrotron frequency is $Q_s^2 \sim -Sh(qeV_o)/(2\pi cp_o)$. The momentum acceptance, dp/p is $\sim(2Q_s/hS)$. With $S = (1/18.7)^2$, $h = 1113$, $Q_s = 780$ Hz and $\omega_o = 0.3$ MHz, $dp/p = 0.16\%$ r.m.s., or a FWHM $\sim 0.38\%$.

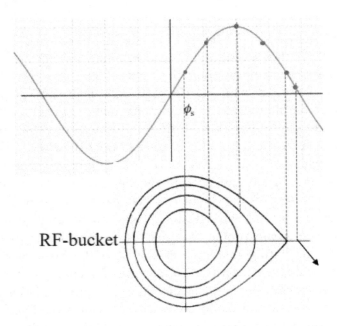

Figure 4.53: Region of stability, the "bucket", under acceleration with synchronous phase ϕ_s. After transition, the maximum stable phase range is approximately π radians with no acceleration where the field increases with time.

Previously, two BPM were used to extract the emittance and dp/p as shown in Figure 4.26. As for chromaticity, the tune shift caused by changing the quadrupole focal length and measuring the resultant tune in BPM was shown in Figure 4.28. The quadrupole kick method was also used to change the tune and find β as in Figures 4.29 and 4.30. The diagnostic use of a change in the r.f. or the B field to find the momentum acceptance is now examined.

If the r.f. frequency is changed by a shift, $\Delta\omega_{rf}$, the synchrotron tune changes as shown above in Eq. (4.32). $\Delta Q_s/Q_s = \Delta p/p = -(1/S)(\Delta\omega_{rf}/\omega_{rf})$. Measuring the change in Q_s with BPM means the $\Delta p/p$ of the bunch can be measured. A change in r.f. changes the central orbit and the centroid of the momentum of the beam therefore shifts. The transverse dispersion, $\eta(s)$, is defined by the shift in position $\Delta x(s) = \eta(s)(\Delta p/p)$. A measurement of the shift enables a measurement of the dispersion, $\eta(s) = -\Delta x(s)S/(\Delta\omega_{rf}/\omega_{rf})$. Knowledge of S is necessary to make the dispersion measurement, however. Some data on the KEK ATF ring using trim steering magnets to correct beam misalignments appears in Figure 4.54. Clearly, the beam must be well aligned prior to measuring the true dispersion since if the dispersion is of order $10\,\mathrm{m}$ and the $\Delta p/p$ is of order 0.1% the shifts will be of order $1\,\mathrm{cm}$.

One possible measurement of $\eta(s)$ would be to measure the dependence of Q_s on V_o since Q_s^2 is proportional to $(ShV_o)/p_o$.

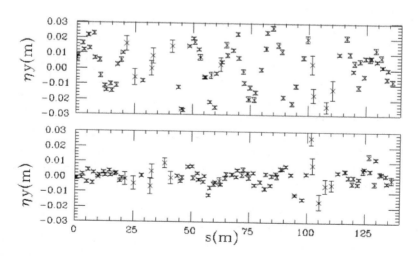

Figure 4.54:   Dispersion, $\eta(y)$ at the KEK ATF damping ring before and after sextupole corrections. There was a $\pm 5\,\mathrm{kHZ}$ ramp of the r.f. leading to a vertical dispersion initially $\sim 2\,\mathrm{cm}$ corrected to $\sim 4\,\mathrm{mm}$.

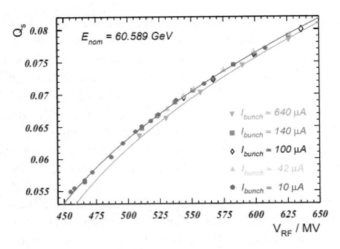

Figure 4.55:   Data from LEP on the tune shift in the synchrotron oscillation frequency as a function of the r.f. accelerating voltage. The fractional shifts have a comparable magnitude and a positive slope.

Data from LEP on the shift in tune as a function of r.f. voltage appears in Figure 4.55. The fractional shifts of Q_s and V_o are comparable, as expected. However, it is easier to make small changes in the r.f. frequency than V_o in general. In this case, $\Delta C/C$ does change. Comparing the shift $\Delta p/p$ or the r.f. shift, $\Delta \omega_{rf}/\omega_{rf}$, to the synchrotron frequency shift, the slope of the linear relationship is S. For the MR with $\omega_{rf} = 330\,\mathrm{MHz}$, the fractional shifts in Q_s and p are, $\Delta Q_s/Q_s = \Delta p/p = -(1/S)(\Delta \omega_{rf}/\omega_{rf})$ and are expected to be ~ 350 times $\Delta \omega_{rf}/\omega_{rf}$. Data from LEP on the tune shift as a function of r.f. are shown in Figure 4.56. The linear relationship between the r.f. shift and the momentum shift is observed and the equal fractional shifts in $\Delta Q_s/Q_s = \Delta p/p$ are represented by the straight line.

Another possible technique to measure the tune shift is to change the dipole power bus while keeping the r.f. constant. This method only changes the beam momentum while keeping the orbit unchanged. Thus, only the effects of quadrupole chromaticity are observed. A measurement of this type is shown in Figure 4.57 where a linear dependence of tune shift on $\Delta B/B$ is observed, as expected.

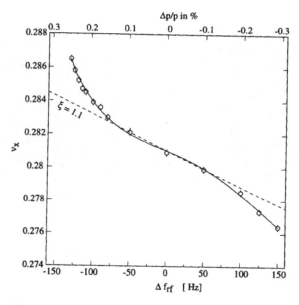

Figure 4.56:   LEP data for a r.f. shift showing the proportional momentum shifts and the tune shifts. The slope is negative, as expected.

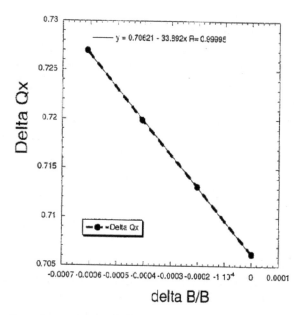

Figure 4.57:   Data from PEP at SLAC on the tune shift as a function of the shift in the dipole field displaying the linear relationship.

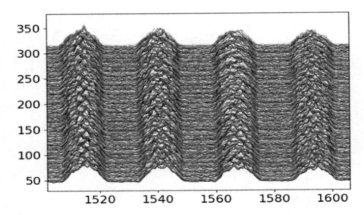

Figure 4.58:   The x axis is time in ns. The y axis is a BPM pickup signal. The traces are separated by 1 turn. The bucket width in phase is $\sim 1/2$ the full cycle or π for a coasting beam.

Yet another possibility is to change the strength of the dipoles and use a fast BPM to measure the period of the stored beam, $\Delta \tau / \tau = -S(\Delta B / B)$. The orbit is unchanged but the centroid beam energy, $\Delta \varepsilon / \varepsilon = \Delta B / B$, is changed.

For a sufficiently long bunch length, a fast response BPM can be used to measure the bunch length in time. The BPM readout can be synchronized to the accelerator r.f. to ensure that the recorded time is locked to the r.f. phase. For the MR, the maximum bucket has acceptance over π radians in r.f. phase, when the beam is not accelerated. In that case, the bucket is 9.4 ns long, or 2.8 m. In the case where dp/p is 0.1%, the bunch is only 1.1 m long. Some data from FNAL appears in Figure 4.58. The coasting beam fills a full bucket of r.f. phase. The FNAL r.f. frequency, f_{rf}, of 53 MHz, with a period of 18.9 ns is evident. The bucket has a spatial extent in z of ~ 2.8 m.

4.18. Strip Lines

For shorter bunch lengths other techniques than BPM or BCT may be needed in order to obtain accurate measurements. Strip lines are an example. They are related to coaxial cables. Strip lines may be thought of as a coaxial cable cut longitudinally and unfolded. The

E and B fields in a coaxial cable are between the inner and outer conductors in the dielectric. For strip lines the fields are more spread out. Strip lines are also often printed on large circuit boards as copper traces on G-10 over the ground plane, for insuring matched impedance signal transmission. Unlike BPM "buttons" strip lines are extended in azimuth with angle α but also extended along a length L of the accelerator z axis, with impedance Z_o. With readout at both ends of the stripline, upper and lower termination R_u, beam current I_b, beam velocity v_b, and strip-line velocity v_{sl}, the signal voltages at the upstream and downstream ends are shown in Eq. (4.40) in the general case.

$$V_u(t) = (\alpha/4\pi)R_u[I_b(t) - I_b(t - L/v_{sl} - L/v_b)]$$
$$V_d(t) = (\alpha/4\pi)R_u[I_b(t - L/v_{sl}) - I_b(t - L/v_b)]$$

(4.40)

If the stripline velocity is equal to the beam velocity, the downstream signal is zero when both ends are matched terminated to Z_{sl} with no reflections. The signal is directional (useful for storage rings) in that only the upstream electrode gives a signal. The pulses are shown schematically in Figure 4.59 and $V_u(t) \sim I_b(t) - I_b(t - 2L/v_{sl})$ while $V_d(t) = 0$. As a special case, if the r.f. bunch length is

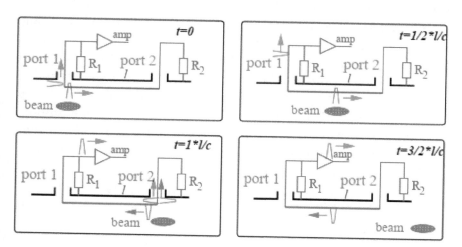

Figure 4.59:   Time development of the pulses in a single stripline if $v_b = v_{sl} = c$ where the downstream pulse is zero. Delta function pulses in time are assumed. The two pulses are spaced in time by $2L/v_b$.

$2L$ in time, Eq. 4.40 shows that there is no upstream signal because the reflected signal cancels it since the beam bunch is in synch with the stripline in this special case.

In the simplest case both beam and strip velocity are effectively c, and the beam has no bunch length but is taken to be a Dirac δ function. In that case, first there is a delta function at the upstream port and then a reflected delta function after a time delay of $(2L)/c$ giving a bipolar upstream pulse and no downstream pulse, as seen in Eq. (4.41) which shows the impedance of the strip line in both the time domain and the frequency domain. The FT of a single delta function is a constant function. In Matlab, the command line input, $fd = \text{fourier}(\text{dirac}(t) - \text{dirac}(t - (2*L)/c))$ leads to the result $1 - \exp(-(L*w*2i)/c)$ where $i = \sqrt{-1}$ for this symbolic utility. Trig identities establish the equivalence of this result with that shown in Eq. (4.41).

$$Z(t) = [\delta(t) - \delta(t - 2L/c)]$$
$$Z(\omega) = j(e^{j\pi/2})(e^{-j\omega L/c})[\sin(\omega L/c)]$$

$$(4.41)$$

For a longer pulse the response is more complex. The FT of a Gaussian in time is a Gaussian in frequency as mentioned previously, $\exp(-t^2/2s^2) \rightarrow \exp(-w^2s^2/2)$. For a beam which is Gaussian distributed in time, but still with beam velocity c and strip-line velocity c, the upstream voltage in time and the impedance in frequency appear in Eq. (4.42). The expression in Eq. (4.42) is a special case of Eq. (4.40). The impedance $Z(\omega)$ is a generalization of Eq. (4.41), and just proportional to the FFT of $V_u(t)$. Using the utility "fourier" one can confirm with a short code snippet that the FT of $\exp[-(t - b)^2/2s^2]$ is $\exp[(-w^2s^2/2 - iwb]$ so that a shift in time creates an additional phase factor in frequency in the FT.

$$V_u(t) = (\alpha/4\pi)Z_{sl}I_b[e^{-t^2/2\sigma^2} - e^{-(t-2L/c)^2/2\sigma^2}]$$
$$Z(\omega) = Z_{sl}(\alpha/4\pi)e^{-(\omega\sigma)^2/2}\sin(\omega L/c)e^{j(\pi/2-\omega L/c)}$$

$$(4.42)$$

When there is a finite beam length in time and a finite beam velocity with a different strip velocity, the upstream voltage is rather

more complicated than Eq. 4.42 for finite beam lengths, as shown in
Eq. 4.43. In addition, the pulses can overlap depending on the speed
of the stripline and that of the beam. This complication will not be
pursued here.

$$V_u(t) \sim e^{-(t+\tau)/2\sigma^2} - e^{(t-\tau)/2\sigma^2}$$

$$\tau = (2L/c)(1/\beta_b + 1\beta_{sl}) \tag{4.43}$$

The deployment of strip-lines in azimuth is similar to that for
BPM. A picture of four strip-lines is shown in Figure 4.60, two
vertical and two horizontal, deployed to yield information on both x
and y of a short beam-bunch. These devices are easiest to use and
interpret if the bunch length is less than the strip length so that the
direct and reflected signals do not overlap and the velocities can be
assumed to be c because the beams are UR.

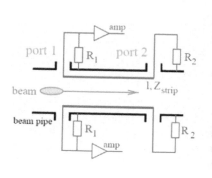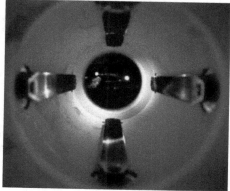

Figure 4.60: A station with four strip-lines, a vertical pair and a horizontal pair.
They are used to supply information on the x and y of the beam centroid for
beams with short bunch lengths.

An App called "StriplineApp" is available to explore more
aspects of a single stripline. The bunch length is a Gaussian in time
with r.m.s. chosen by "Slider" with a range available from very short,
to the case of overlapping direct and reflected signals. The stripline
output in both the time domain and the frequency domain is plotted.
The stripline length is 30 cm or 1 ns. The frequencies extend up to

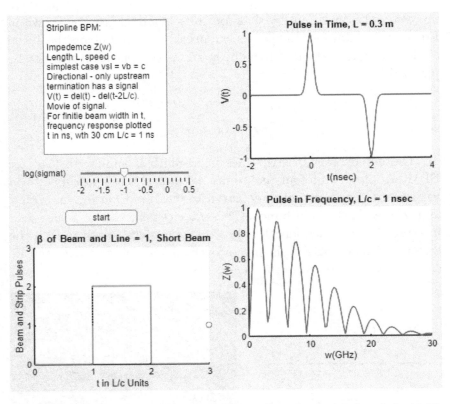

Figure 4.61: Output of $V(t)$ and $Z(\omega)$ for a Gaussian bunch length in t with r.m.s. chosen by Slider. The movie is a schematic representation of a Dirac delta function bunch crossing the stripline which has $L/c = 1\,\text{ns}$.

tens of GHz. A "movie" is shown schematically which indicates a delta function in time for the beam and the reflected pulses. Output of the App appears in Figure 4.61. The FT plot displays the modulus of the impedance rather than the real part.

4.19. Synchrotron Radiation — γ

In the case of extremely relativistic beams, for example from electron accelerators, additional techniques to interrogate the beam properties become available. To begin, the Lorentz transformation for photons defines the relationship between polar angles in different reference frames independent of any dynamics. This is purely kinematics. The forward emission of photons is called the "search light" effect. It is a general effect due to SR. The transformation for photon energy and momentum implies a simple relationship between photon angles for an isotropic distribution in one frame, the $*$ frame, transformed to a distribution in a frame moving with velocity βc. The relationships are: $p = \gamma p_*[1 + \beta \cos(\theta_*)]$ and $p\cos(\theta) = \gamma p_*[\cos(\theta_*) + \beta]$. For reference frames in extreme relative motion an isotropic photon angular distribution becomes sharply peaked in the direction of motion of the reference frame.

$$d(\cos\theta) = (d\cos\theta_*)/[\gamma^2(1 - \beta\cos\theta_*)^2] \qquad (4.44)$$

The transformation of photon angles is explored in the App "SearchLight_App". The Lorentz transformations are quoted above and in the script and the distribution of angle is solved symbolically using "diff". The maximum angle is found using the utility "max" and the value is $\sim 1/\gamma$. A typical output of the App, where γ is chosen by "Slider" appears in Figure 4.62. The user can check that choosing $\gamma = 1$ gives an isotropic distribution.

NR radiation by an accelerated charge has an angular distribution in lowest multipole order which is dipole, with emission preferentially perpendicular to the acceleration (Larmor). In a magnetic field, the Lorentz force equation implies a rotation of the charged particle which has the instantaneous circular frequency, $\omega_o = qeB/(\gamma m) = \beta c/\rho \sim c/\rho$. The radiated power (NR) was presented earlier in the discussion of NR radiation in Eq. (3.25). As quoted before, the charge is qe and the acceleration (NR) is a, as in Section 3.9. Other forms for the radiated power can be found using $\omega_o = \beta c/\rho$, $a = c^2/\rho$ or $\tau = 2\pi\rho/c$, specifically for circular motion.

$$\langle P \rangle = d\langle U \rangle/dt = (qea)^2/(6\pi\varepsilon_o c^3) = (2/3)q^2\alpha\hbar a^2/c^2 \qquad (4.45)$$

In the UR case, radiation is thrown forward along the direction of motion (search light). For circular motion with velocity β, along z, perpendicular to $d\beta/dt$, along x instantaneously, the radiated energy per solid angle and frequency is, in small angle approximation,

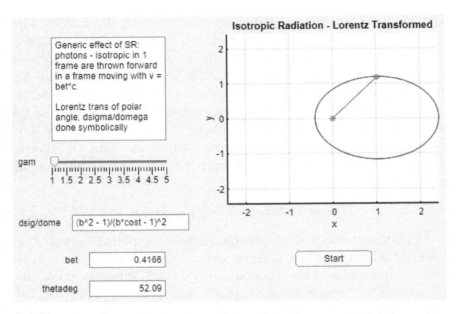

Figure 4.62:   The angular distribution of isotropic photons moving with a Lorentz factor γ. The maximum angle is found, shown as a red line and displayed in an "EditField". The photons can always go backward.

displayed in Eq. (4.46). The dimensionless scaled variables y and s are used, where both variables are of order one. The small angle and UR approximation for $(1 - \beta \cos \theta)$ makes the search light effect explicit. Matlab supplies the modified Bessel functions, $K_\nu(s)$, as it does for most special functions, which makes computations simple. Using scaled variables, the double differential cross-section is here dimensionless and the radiative coupling factor α appears as it did for Cerenkov emission, since both involve a single vertex for real photon emission. Later, in Eqs. (4.49) and (4.50) the full dimensional factors are included

$$(1 - \beta \cos \theta) \sim (\theta^2 + 1/\gamma^2)/2$$
$$d^2 \langle U \rangle / d\Omega d(\hbar\omega) = (\alpha/4\pi^2)(\omega/\omega_o)^2 (2/3\gamma^2)^2 (1 + y^2)^2$$
$$\{K_{2/3}^2(s) + [y^2/(1 + y^2)]K_{1/3}^2(s)\} \tag{4.46}$$
$$y = (\theta\gamma), s = (\omega/3\omega_o\gamma^3)(1 + y^2)^{3/2} = (\omega/3\omega_c)(1 + y^2)^{3/2}$$

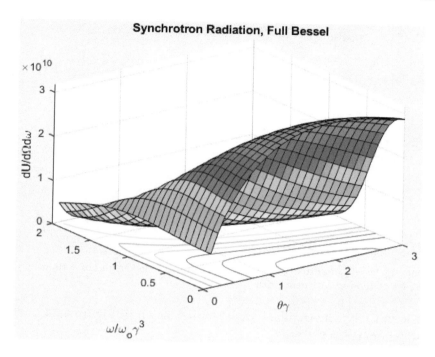

Synchrotron Radiation, Full Bessel

Figure 4.63:   Surface of $d^2U/d\Omega d\omega$ as a function of y and ω/ω_c with contours.

The doubly differential cross section is displayed as a surface in (y,s) space in Figure 4.63 where the parameters, ω_o and γ, are set to be those of the LEP collider at CERN with 50–50 GeV electron–positron collisions. The script "Synch_dU_domega_dw" uses the utility "surfc". There are correlations between the angle and the frequency of radiation. Note that the angular distribution is more peaked at large values of s than at smaller emitted frequencies. At all angles there is a steep falloff of energy with frequencies beyond a cutoff frequency ω_c, defined to be $\omega_o\gamma^3$ which appears in the scaled variable s.

The differential power into unit solid angle is found by integration over s, and can be solved for analytically without the small angle approximation. It is shown in Eq. (4.47). The acceleration here is taken to be a $= (\beta c)^2/\rho$ using the NR centrifugal force in the case of circular motion. In general, in SR the acceleration is quite a complex quantity, the discussion of which is avoided here. A plot is made using

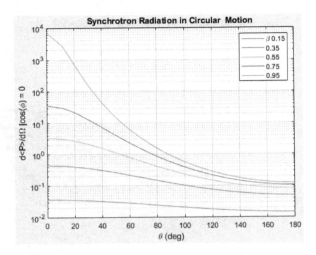

Figure 4.64: Angular distribution, $d\langle P\rangle/d\Omega$ for different values of β showing the increase of the forward peak with β.

the script "Synch_Angular" with results shown in Figure 4.64 in the case where $\cos\phi = 0$.

$$a_1 = [q^2\alpha\hbar/(4\pi c^2)][a^2/(1-\beta\cos\theta)^3]$$
$$d\langle P\rangle/d\Omega = a_1\{1 - [\sin^2\theta\cos^2\phi]/[\gamma^2(1-\beta\cos\theta)^2]\}$$

(4.47)

The total radiated power is the NR result, Eq. (4.35), times γ^4. The radiated power can be related to the Thomson cross-section, σ_T and the energy density in the magnetic field, $u_B = B^2/(2\mu_o)$, which is accelerating the electrons. In an NR case, related to synchrotron radiation, the scattered power for Thomson scattering is just $\sigma_T(cu_{\text{in}})$ where u_{in} is the incident photon energy density. For the UR case the energy density is that of the magnetic field.

$$\langle P\rangle = (4/3)\sigma_T(cu_B)(\beta\gamma)^2$$

(4.48)

This compact and instructive result can be cast into a second form, more conventional but perhaps less intuitive, using the expressions for the Thomson cross-section, the rotation frequency, and the radius of curvature. For circular motion, taking $a = c(d\beta/dt) = v^2/\rho$, which is not generally correct in SR, the total power is the integral of Eq. (4.47) and is just $(\beta\gamma)^4$ times the NR Larmor expression with $q = 1$:

$$\langle P \rangle = e^2 c (\beta\gamma)^4 / (6\pi\varepsilon_o \rho^2) = (2/3)\alpha\hbar(c/\rho)^2 (\beta\gamma)^4 \qquad (4.49)$$

The radiated photon energies extend up to a frequency which is approximately ω_c, much larger than the rotation frequency. The emission angle is $\sim 1/\gamma$ at the highest frequencies, as expected from the search light effect. More generally, $\theta_c = (1/\gamma)(\omega_c/\omega_o)^{1/3}$ and higher energy photons have smaller critical angles as seen in Figure 4.63. The typical maximum radiated photon energy in the UR approximation is simply $\hbar\omega_c$. To convert from power, P to ΔU, the energy loss per turn at UR energies, is $\Delta U = \langle P \rangle \tau$. The total beam power lost is the energy loss per turn per particle times the beam current in A. Approximating UR behavior with $\beta = 1$;

$$\langle P \rangle = (2/3)q^2 \alpha\hbar\omega_o^2 \gamma^4, \tau = 2\pi/\omega_o$$
$$\Delta U = (4\pi\alpha q^2 \hbar\omega_o/3)\gamma^4 \qquad (4.50)$$

Numerical values for energy loss per turn, ΔU, per electron, $q = 1$, in keV, are shown in Eq. (4.51). The radiated power sets a practical limit for an electron storage ring. The LEP collider at CERN was the largest circular machine ever built and its energy was limited by the available r.f. power that could be installed in the LEP ring. The only way to increase the energy at fixed r.f. power is to increase the radius of the ring.

$$\Delta U = (qe)^2 \gamma^4 / (3\varepsilon_o \rho)$$
$$\Delta U (\text{keV}) = 88.9\varepsilon^4 (\text{GeV})/\rho(m) \qquad (4.51)$$
$$\hbar\omega_c(\text{keV}) = 2.2\varepsilon^3 (\text{GeV})/\rho(m) = 0.665\varepsilon^2 (\text{GeV})B(T)$$

The numerical estimates for a circular electron machine are set out in the App "Synch_Circ_LEP". The electron energy is set by "Slider" as is the radius of the machine. The initial values of both are set to approximate the LEP machine working at the Z boson mass. The output of the App in that case is shown in Figure 4.65. The magnetic field, the photon cutoff energy, the energy loss per turn and the approximate number of radiated photons per turn are displayed using numerical "EditFields". The user can explore the

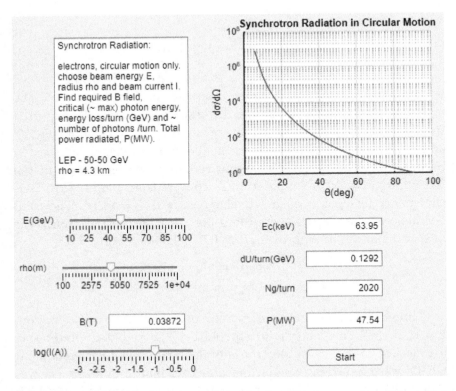

Figure 4.65: Radiation loss in a circular electron storage ring. The initial parameters are the settings which are approximately those of LEP producing Z bosons on resonance.

tradeoff of radius, B field, and energy loss and see that the γ^4 factor dominates the optimization. A "Slider" chooses the electron beam current and therefore the power loss. Assuming a limit of ~ 100 MW due to costs and available space for r.f. cavities, the user can see what combinations of current and energy obey such a limit.

The sharp angular emission of synchrotron radiation can be used to find the emittance of an e-beam. For UR e-beams the technique is in common use. In e-machines the γ factor can be very large, leading to measurable radiation due to acceleration (bending magnets or specialized magnets — wigglers, undulators). If the wavelength of the radiation is less than the bunch length the radiation is incoherent and is proportional to the number of particles in the bunch, N. If not, it is coherent, and the radiated power goes as N^2. As a numerical

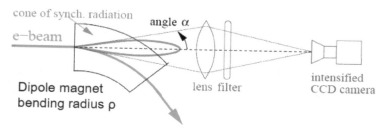

Figure 4.66:   Schematic of the use of a dipole magnet to study the transverse beam distribution using synchrotron radiation.

Figure 4.67:   PEPII transverse image using synchrotron photons. With a gated CCD camera one can also make bunch-by-bunch measures of the transverse beam size.

example, a $L = 6$m, 2T dipole traversed by a 100-GeV electron has a radius of curvature of 166 m. The total bend angle is about 0.036 rad which results in the emission of about 24 photons per electron with a critical energy of 13.2 MeV. The total energy loss is about 0.31 GeV or about 0.31%. A schematic of the setup to measure the emitted photons is shown in Figure 4.66 while data from PEPII at SLAC is shown in Figure 4.67.

4.20. Laser Backscatter — ε_o

Compton scattering has already been discussed in Section 3.10. In particular, Eq. (3.23) gives the relationship between the outgoing photon energy and the photon scattering angle. For backward photon scattering, which is always possible, the energy ratio of final and initial photon energies is ~1-twice the initial photon energy divided by the electron rest energy. We now consider the kinematics in the special case of an electron moving along the z axis with velocity βc, and a photon, with frequency ω_o, incident at a polar angle θ_o. The photon final state has frequency ω and polar angle θ. The kinematics is, as always, just the conservation of SR momentum and energy and is shown in Eq. (4.52) in the approximation that the incident photon energy (laser) is small with respect to the electron beam energy. The three constraint equations are used to remove the scattered electron variables from Eq. (4.52). In what follows, the special case of a head-on collision is chosen, $\cos \theta_o = -1$, in which case the final state photon has a maximum energy when back scattered, of $4\varepsilon_o\gamma^2$. It is assumed that the electron is UR. For example, with a 5.1-GeV electron beam and a 0.1-eV laser, the maximum recoil photon is a hard X-ray with 40 MeV energy. For collisions which are not head-on the maximum recoil photon energy is less.

$$\omega/\omega_o = (1 - \beta \cos \theta_o)/(1 - \beta \cos \theta)$$

$$\beta \sim 1 - 1/(2\gamma^2), (1 - \beta \cos \theta) \sim (1/2\gamma^2)[1 + (\theta\gamma)^2]$$

$$\varepsilon_{\max} = 4\varepsilon_o\gamma^2 \sin^2(\theta_o/2) \qquad (4.52)$$

$$\omega/\omega_o \to (1 + \beta)/(1 - \beta \cos \theta) \sim 4\gamma^2/[1 + (\theta\gamma)^2]$$

The cross-section for the process is shown in Figure 4.68 plotted as a function of $4\varepsilon\varepsilon_o/(m_ec^2)^2 = x$ using the script

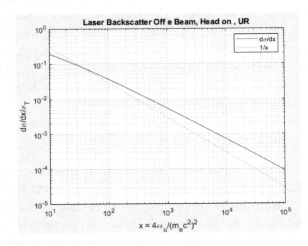

Figure 4.68: Cross-section with respect to σ_T for head-on laser backscattering as a function of the laser energy times the electron beam energy scaled to the electron rest energy squared, or x. The dotted line exhibits the leading $1/x$ behavior.

"cross-section_e_laser_scat", where ε is the incident electron energy. The cross-section is scaled to the related Thomson cross-section and falls with x as an approximate power law with a power between -1 and $-1/2$. The lowest order terms in the cross section are, $(d\sigma/dx)/\sigma_T \sim (3/8x)$. The full expression for the cross-section is not particularly illuminating and can be found in the script.

As mentioned previously electromagnetic processes often have a $1/x$ type scaling. The "Compton edge" at the maximum scattered photon energy can then be used to precisely determine the central momentum of an electron beam. In Feynman diagram language the Compton process has an "s channel" diagram but also a "t channel" process with virtual electron exchange. The latter diagram leads to a sharp backward photon peak at the highest photon energy.

A laser beam has an energy, ε_o, typically of $\sim$ eV. The scattered photons have substantially higher energy, Eq. (4.52). The backscattered surface of ω and θ of the outgoing photon is plotted as a function of the incident electron γ factor in the script "laser_e_backscat" and displayed in Figure 4.69 where the full UR

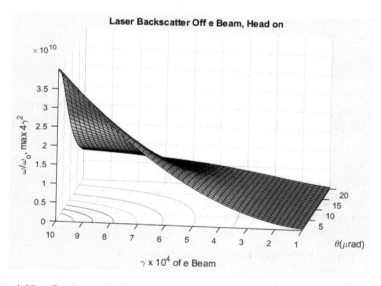

Figure 4.69: Surface of the backscattered photon energy as a function of backscattered angle and electron beam energy. The maximum recoil energy is only weakly dependent on angle.

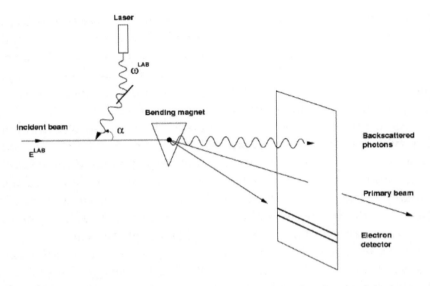

Figure 4.70: Schematic diagram of setup to measure the electron beam energy using the kinematics of Compton backscattering. In this case both the unscattered laser beam, the scattered electron, and the backscattered photon are detected which makes for tighter kinematic constraints.

approximation is made, $\omega/\omega_o \sim 4\gamma^2/[1 + (\gamma\theta)^2]$. The surface is very sharp in angle and rises rapidly with electron energy.

A schematic diagram of a setup to examine the backscattered laser photons appears in Figure 4.70. In this case the laser does not impinge on the beam head-on. Some actual backscattered data is shown in Figure 4.71. The sharp Compton edge is evident. Indeed, the edge is smeared by the momentum spread in the beam and by the experimental resolution of the scattered photon energy. The transverse beam size can also be probed using a very collimated laser beam. In extreme cases laser pulses of picoseconds length can be used to interrogate the bunch length.

4.21. Optical Transition Radiation — γ

Previously, in Section 3.9, transition radiation (TR) was discussed in the context of detecting UR single electrons in a beam-line. Since only about α photons of high energy are emitted per particle per

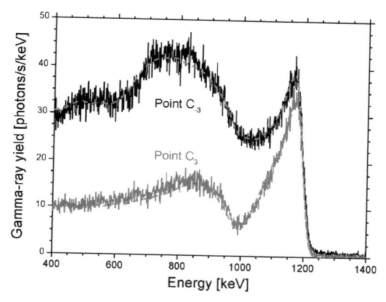

Figure 4.71:   Data on the energy spectrum of backscattered photons. The photon energy displays the maximum energy with a sharp "Compton Edge" of a few keV out of 1.2 MeV.

transition, many foils are needed to achieve good detection efficiency. The detected objects are X-rays. Since they are not heavily absorbed by the stack of thin radiating foils and the stack is approximately transparent to its own emissions. In the case of an electron beam in an accelerator, these constraints are not relevant. Use is then made of optical TR photons to explore the electron beam properties. As seen in Figure 3.23, the TR frequency spectrum peaks at low frequencies as is usual for radiative processes. In that case, optical photons are more copious and are also easier to work with. The use of light rather than X-rays simplifies the instrumentation. Indeed, it is common to insert a thin screen into the beam and look at the optical TR (OTR) which is emitted both forward and backward. Clearly, this technique is not a fully non-destructive one and the beam will be perturbed. A schematic view of the setup is shown in Figure 4.72. Typically, the screen is tilted at 45° in which case the OTR is emitted near the vertical axis. This layout differs from that of Figure 3.24 with normal

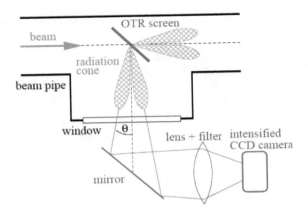

Figure 4.72:   Layout of OTR readout with a thin screen inclined at an angle, an optical window, and the lens for focusing the beam image onto a CCD camera.

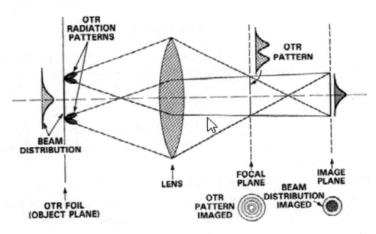

Figure 4.73:   Schematic of the pattern using OTR for beam diagnostics in assessing the transverse size of the beam.

incidence. The backward photons are then easy to observe and the resulting "doughnut" yields information on the beam profile.

An example of the use of OTR for transverse beam diagnostics appears in Figure 4.73. The forward OTR pattern is a toroidal doughnut while the beam pattern is a Gaussian distribution in transverse position. The beam size can even be examined turn by turn, as shown in Figure 4.74.

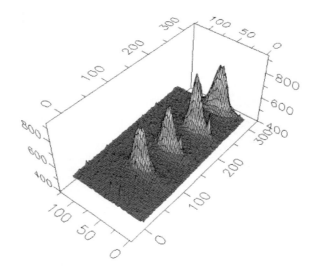

Figure 4.74:   Turn by turn OTR data from the SPS at CERN.

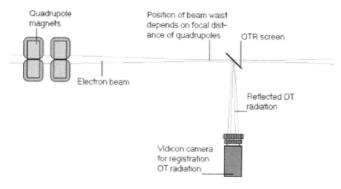

Figure 4.75:   Schematic view of the use of a quadrupole "bump" to explore the electron beam emittance with an OTR dataset.

OTR is also used for electron beam diagnostics using the deflection of magnets rather than using a screen. For example, with a quadrupole "bump" the beam emittance ellipse can be mapped out similarly to the method mentioned previously with BPM. A schematic view of a quadrupole plus OTR screen is shown in Figure 4.75. Using a dipole instead enables the measurement of the beam momentum spread.

Chapter 5

Summary

This book has attempted to cover the physics of the instrumentation used in particle detectors, particle beams and in accelerators. The physics has not been rigorous, but rather intuitive. Great use has been made of the Matlab suite of software utilities. Apps have been written that aim to give the user an experience with how a specific instrument operates with different parameter values of its operation chosen by the user.

There was an initial section on Matlab tools. However, much more material is available in tutorials supplied by Matlab and in the very extensive Matlab documentation. The text itself introduces specific utilities whenever the subject at hand uses a particular utility in an App. The concept which was followed was to discuss a tool when it is needed and when its use can then be examined in a specific App.

Chapter 2 introduced the very basic physics language which is needed to start to explore the topics to be discussed in the sections on instrumentation. It was meant only to define the most basic physics concepts and parameters which were needed to get started.

Chapter 3 explored both beam and detector instrumentation. There was a short introduction to some of the basic physics concepts. Then the more specific physics was introduced when it was necessary to understand how a particular tool functioned. The same strategy was adopted in the fourth section on accelerators. Specific physics topics appear only when necessary after an introduction which defines the key concepts.

The full complexity of the many instruments was not discussed since the required length of such a text would be enormous. Rather, a more schematic approach to a class of instruments was adopted. For example, the "accelerator" used to explain many of the concepts was simply 96 sequential FODO elements. No attempt was made to explore injection, extraction, special insertions, or collision regions even though those are some of the most important aspects of a functional accelerator.

Clearly the quite superficial coverage of the topics in this text is meant to invite the reader to explore several possible jumping off points. They could study much more rigorous textbooks covering some specific physics topic of interest. Or more specific technical details of particular instruments could be explored. Finally, the many topics related to instrumentation used for processing the signals "downstream" of the front-end signal formation discussed here could be studied. Examples might include electronics, logic gate arrays, triggering of an experiment, data acquisition, data storage and analysis. It is to be hoped that the reader will go on well beyond the limitations of this text.

Appendix A: Matlab Scripts

A.1 nuc_lam_A.m

```
%
close all
clear all
%
A = [238 207 184 63.5 40 27 20 9];
%lam = [114.1 118.6 110.4 84.2 75.7 69.7 65.7 55.3]; % elastic + inelastic
lam = [209 199.6 191.9 137.3 119.7 107.2 99.0 77.8]; % inelastic only
lam = lam .10; % gm/cm^2 -> kg/m^2
% U Pb W Cu Ar Al Ne Be
%
AA = linspace(9,238);
llam = 350 .(AA .^0.3333 );
loglog(A,lam,'o',AA,llam,'-r');
%
grid
title('Nuclear Interaction Length vs A')
xlabel('A')
ylabel('\lambda_I(kg/m^2)')
%
```

A.2 Uniform_E_App

```
%
% trajectory of a charge in a uniform E field - symbolic
%
syms a t z zp b bc zc q Eo m
%
% dp/dt = qEo, p = (qEo)t, beta = p/E = at/√((at)^2 + 1), a =
qEo/m \n ')
%
p = qEot; % beta follows since p = bgm
b = at/√((at)^2+1);
z = int(at/√((at)^2+1));
z = z - 1/a; % z(0) = 0
pretty(z)
app.ptEditField.Value = char(p);
app.betaEditField.Value = char(b);
app.zEditField.Value = char(z);
%
% Classical Non-relativistic Results \n ')
%
bc = at ;        % NR classical values
zc = (att)/2;
%
tt = linspace(0,5,20);
%
a = 1;
%
```

```
t = tt;
bb = eval(b); % dzdt
zz = eval(z); % distance
bbc = eval(bc); % NR classical limits
zzc = eval(zc);
%
for i = 1:length(tt)
    plot(app.UIAxes,tt(i),bb(i),'-bo',tt(i),bbc(i),'r:')
    title(app.UIAxes,'\beta in a Uniform Electric Field ')
    xlabel(app.UIAxes,'ct')
    ylabel(app.UIAxes,'\beta')
    legend(app.UIAxes,'SR','Classical','Location','Northwest')
    pause(0.2)
    xlim(app.UIAxes,[0 5])
    ylim(app.UIAxes,[0 5])
    hold(app.UIAxes, 'on')
end
hold(app.UIAxes, 'off')
%
for i = 1:length(tt)
    plot(app.UIAxes2,tt(i),zz(i),'-bo',tt(i),zzc(i),'r:')
    title(app.UIAxes2,'z as a function of t')
    xlabel(app.UIAxes2,'ct')
    ylabel(app.UIAxes2,'z')
    legend(app.UIAxes2,'SR','Classical','Location','Northwest')
    pause(0.2)
    xlim(app.UIAxes2,[0 5])
    ylim(app.UIAxes2,[0 14])
    hold(app.UIAxes2, 'on')
end
 hold(app.UIAxes2, 'off')
%
        end
    end
```

A.3 Uniform_B_App

```
function update(app)
        %
        % Charged particle in a constant B field, Bo, along y
        %
        syms t ax(t) az(t) r x(t) z(t) axs azs
        %
        % dp/dt = q(v x B), p constant, radius of curvature r constant
        % initially moving along z axis, direction cosines a
        %
        ode1 = diff(ax(t)) == -az(t)/r;
        ode2 = diff(az(t)) == ax(t)/r;
        odes = [ode1; ode2];
        %
        init = [ax(0) == 0; az(0) == 1];
        [axs(t),azs(t)] = dsolve(odes,init);
```

```
%
app.axEditField.Value = char(simplify(axs));
app.azEditField.Value = char(simplify(azs));
%
%Positions x and z ;
%
x = int(axs);
z = int(azs);
app.xEditField.Value = char(x);
app.zEditField.Value = char(z);
%
% now numerical plots
% Take a step of arc length s in a uniform magnetic field of
magnitude Bo
% along the y axis, charge qe, incident position,xo,yo, momentum
components
% pox, poy poz - exact solutions. Movie made
%
Bo = 10 .app.BTSlider.Value; % field in KG, slider in T
%
ss = 60; % arc length
po = [ 1 1 0]; % movng initially in x y
xo = [0 0 0];
ptot = √ (po(1) .^2 + po(2) .^2 + po(3) .^2);
s = linspace(0 ,ss, 50); % arc length
%
for i = 1:length(s)
    %
    % no force along y
    %
    yyy(i) = xo(2) + (s(i) .po(2)) ./ptot; % straight in (y,s) plane
    %
    % find radius of curvature (signed), and bend angle, Kg, GeV, m,
    e
    %
    a = 33.356; % using q = proton charge, MKS
    rho = (a .ptot) ./Bo; % radius of curvature
    phi = s(i) ./rho; % bend angle
    %
    % direction cosines rotate by phi, positions correlated
    %
    cp = cos(phi);
    sp = sin(phi);
    p(1) = po(1) .cp + po(3) .sp;
    p(3) = po(3) .cp - po(1) .sp;
    p(2) = po(2);
    xxx(i) = xo(1) - (rho .(p(3) - po(3))) ./ptot;
    zzz(i) = xo(3) + (rho .(p(1) - po(1))) ./ptot;
    %
    plot3(app.UIAxes, xxx(i) , yyy(i), zzz(i),'-bo')
    title(app.UIAxes,'Three Dimensional Orbit in a By Field')
    xlabel(app.UIAxes,'x');
    ylabel(app.UIAxes,'y');
    zlabel(app.UIAxes,'z');
```

```
        pause(.3)
        xlim(app.UIAxes,[-4 4])
        ylim(app.UIAxes,[0 60])
        zlim(app.UIAxes,[-8 0])
        hold(app.UIAxes,'on')
    end
    %
    plot3(app.UIAxes,xxx , yyy, zzz,'-b')
    hold(app.UIAxes,'off')
    %

End
```

A.4 Cyclotron

```
function update(app)
        %
        % Look at Cyclotron Operation - NR only - p or ion
        % work in dimensionless units as possible
        %
        % Cyclotron w = qB/m, r = mv/qB = v/w, E = (qBr)^2/2m
        % Relativistic Effects, m -> E, w Decreases and r Increases by gamma
        %
        n = app.Nturns2Slider.Value;
        dE = app.d_KE_deesSlider.Value;
        %
        wt = linspace(0.0, pi .n,400);
        % initial radius
        a(1) = 0.05;   % initial energy/radius - p source assumed
        %
        xx(1) = a(1); % starting up
        yy(1) = 0;
        dwt = wt(2)- wt(1); % equal time step movie
        imax = length(wt);
        for i = 2:imax
            j = round(1 +(i n ./imax));  % which half rotation?
            E(j) = j .dE; % kinetic energy gains dE for each i/2 rotation
            a(j) = √(E(j)); % dE ~ rdr, radius for this 1/2 rotation
            xx(i) = xx(i-1) + a(j) .cos(wt(i)) .dwt;
            yy(i) = yy(i-1) + a(j) .sin(wt(i)) .dwt;
            if √(xx(i) .^2 + yy(i) .^2) > 5 ;  % set radius of the Dees
                imax = i;
                break
            end
        end
        %
        % draw "dees"
        %
        xdee = linspace(-5,5);
        ydee = √(5.0 .^2 - xdee .^2);
        yp = [0.1 0.1];   ym = [- 0.1 -0.1];
        xp = [-5.0 5.0]; xm = [-5.0 5.0];
```

```
    %
    for i = 1:imax
        plot(app.UIAxes,xdee,ydee,'b',xdee,-ydee,'r',xp,yp,'b-',xm,ym,'r-
        ',xx(i),yy(i),'og')
        xlabel(app.UIAxes,'x(m)')
        ylabel(app.UIAxes,'y(m)')
        title(app.UIAxes,'Cyclotron Orbits - Non-Relatisivtic')
        xlim(app.UIAxes,[-5 5])
        ylim(app.UIAxes,[-5 5])
        pause(0.02)
    end
    %
    plot(app.UIAxes,xdee,ydee,'b',xdee,-ydee,'r',xp,yp,'b-',xm,ym,'r-
    ',xx,yy,'-g')
    xlabel(app.UIAxes,'x(m)')
    ylabel(app.UIAxes,'y(m)')
    title(app.UIAxes,'Cyclotron Orbits - Non-Relatisivtic')
    xlim(app.UIAxes,[-5 5])
    ylim(app.UIAxes,[-5 5])
    %
end
%
```

A.5 ExB_App

```
function update(app)
    %
    % Charged particle in constant E and B Fields
    % crossed E and B, Eo along y, Bo along x, beam initially (0,0,0)
    %  initial velocity along z, accel = (e/m) (E + v x B) - NR
    %  dp/dt in SR case, p = g  b  m c
    %
    global Bx Ey
    %
    Ey =  1e6 .app.EoMVmSlider.Value; % electric field in MV/m
    %
    % B field large = 1 T to see curvature
    % B fielld = 0.02 T appropriate to a ~ 1 GeV KEK Kaon beam
    val = app.BoDropDown.Value;
    tf = strcmpi(val,'1T');
    if tf == 1
        Bx = 1;
    end
    tf = strcmpi(val,'0.02T');
    if tf == 1
        Bx = 0.02;
    end
    % Magnetic Field Bz in T;
    boz = app.bozSlider.Value; % Initial beta;
    c = 3e8;
    betao = Ey ./(c .Bx) ; % drift velocity
    app.b_driftEditField.Value = betao;
    %
```

```
tspan = linspace(0,10);    % equal steps in ctime to makes movies, 10
m
[t,y] = ode45(@app.EB,tspan,[0 ; 0 ; boz ; 0 ; 0 ; 0]); %  initial
conditions
%
% dbetax = y(1), dbetay = y(2), dbetaz = y(3); dx/dct = betx
%
plot(app.UIAxes,t,y(:,2),'r-',t,y(:,3),'b:')
title(app.UIAxes,'\beta in E x B Field')
xlabel(app.UIAxes,'ct(m)')
ylabel(app.UIAxes,'\betay, \betaz')
legend(app.UIAxes, '\betay','\betaz','Location','southeast')
%
plot(app.UIAxes2,t,y(:,5),'r-',t,y(:,6),'b:')
title(app.UIAxes2,'position in E x B Field')
xlabel(app.UIAxes2,'ct(m)')
ylabel(app.UIAxes2,'y,z (m)')
legend(app.UIAxes2,' y','z','Location','southeast')
%


end

function dydt = EB(app,t,y)
    % --------------------------------------------------------------------
    --------
    %
    global Bx Ey
    % d1 = (eEo)/(mc^2), d2 = (cBo)/Eo = 1/betao
    e = 1.6e-19; m = 1.67e-27; c = 3e8; % p charge, mass
    d1 = (e .Ey) ./(m .c .^2); d2 = (c .Bx) ./Ey ;
    gami = √(1.0 - y(1) .^2 - y(2) .^2 -y(3) .^2);
    d1 = d1 .gami;
    dydt = zeros(6,1);
    dydt = [-d1 .y(1) .y(2) ; d1 .(1.0 + d2 .y(3)-y(2) .^2);  d1 .(-
    d2 .y(2)-y(3) .y(2)) ; y(1) ; y(2) ; y(3)];
    %
End
```

A.6 Quad_pot_field.m

```
%
close all
clear all
%
syms pot x y ph dBdr f Bx By L q p k MF MD
%
pot = dBdrxy; % quad potential
Bx = -diff(pot,x); % field = gradient(potential)
By = -diff(pot,y);
%
% phase, focal length - thin lens f =1/kL
%k = qdBdr/p
%ph = √(k)L
```

```
%
MF = [cos(ph), sin(ph)/√(k); -sin(ph)√(k), cos(ph)]
MD = [cosh(ph), sinh(ph)/√(k); sinh(ph)√(k), cosh(ph)]
%
[xx,yy] = meshgrid(-1:0.05:1,-1:0.05:1);
potent = xx .yy;
[BBx,BBy] = gradient(potent,0.05,0.05);
contour(xx,yy,potent)
hold on
quiver(xx,yy,BBx,BBy), hold off
title('Quadrupole Potential and Fields')
xlabel('x')
ylabel('y')
%
```

A.7 Doublet_Thin_Symbolic

```
function update(app)
          %
          % thin lense doublet - symbolic
          %
          ddo = app.dQ1Slider.Value ; % Distance start to Q1 CL
          dd = app.dQ1Q2Slider.Value ; % Distance Q1 to Q2 CL
          ddc = app.dQ2FocSlider.Value ; % Distance Q2 CL to constraint
          %
          % f1 is first F x quad, f2 is second D x, do is initial to f1,
          % d is f2-f1 , dc is f2 to constraint
          %
          % symbolic thin lens doublet - point to par, par to point, point to
          point
          %
          syms f1 f2 do d dc M1 M2 M3 M4 M5 MT N1 N2 N3 N4 N5 NT
          syms pt par x1 x2 x3 x4 x5 y1 y2 y3 y4 y5
          %
          M1 = [1 do ; 0 1]; M3 = [1 d ; 0 1]; M5 = [ 1 dc ; 0 1]; % drifts
          M2 = [1 0; -1/f1 1]; M4 = [1 0; 1/f2 1]; % quads
          MT = M5M4M3M2M1;
          %
          % now the (y,z) view
          %
          N1 = [1 do ; 0 1]; N3 = [1 d; 0 1]; N5 = [ 1 dc; 0 1];
          N2 = [1 0; 1/f1 1]; N4 = [1 0; -1/f2 1];
          NT = N5N4N3N2N1;
          fprintf(' The (x,z) and (y,z) doublet matrices \n');
          simplify(MT);
          pretty(MT);
          simplify(NT);
          pretty(NT);
          %
          % now look at point to parallel, point to point or parallel to point
          % pt->par M22 = 0, pt->pt M12 = 0, par->pt M11 = 0
          %
          val = app.ConstraintDropDown.Value;
          tf = strcmpi(val,'->||');
          if tf == 1
              eqs = [MT(2,2)==0,NT(2,2)==0]; % point to parallel
```

```
        pin = [0 ; 1];
    end
    tf = strcmpi(val,'->');
    if tf == 1
        eqs = [MT(1,2)==0,NT(1,2)==0]; % point to point
        pin = [0 ; 1];
    end
    tf = strcmpi(val,'||->') ;
    if tf == 1
        eqs = [MT(1,1)==0,NT(1,1)==0]; % parallel to point
        pin = [1 ; 0];
    end
    vars = [f1 f2];
    [solf1,solf2] = solve(eqs, vars); % symbolic solution
    fprintf(' Symbolic Solution for f1 , f2 \n');
    f1 = solf1(1);
    f2 = solf2(1);
    sf1 = simplify(f1);
    sf2 = simplify(f2);
    app.f1EditField.Value = char(sf1); % symbolic
    app.f2EditField.Value = char(sf2);
    %
    % numerical evaluations
    %
    do = ddo;
    d = dd;
    dc = ddc;
    FF1 = eval(f1);
    FF2 = eval(f2);
    app.f1EditField_2.Value = FF1; % numeric
    app.f2EditField_2.Value = FF2;
    %
    M1 = [1 do ; 0 1]; M3 = [1 d; 0 1]; M5 = [ 1 dc; 0 1];
    M2 = [1 0; -1/FF1 1]; M4 = [1 0; 1/FF2 1];
    N1 = [1 do ; 0 1]; N3 = [1 d; 0 1]; N5 = [ 1 dc; 0 1];
    N2 = [1 0; 1/FF1 1]; N4 = [1 0; -1/FF2 1];
    X1=M1pin; X2=M2X1; X3=M3X2; X4 =M4X3; X5=M5X4;
    Y1=N1pin; Y2=N2Y1; Y3=N3Y2; Y4 =N4Y3; Y5=N5Y4;
    zdoub = [0,do,do,do+d,do+d,do+d+dc];
    plot(app.UIAxes, zdoub, [pin(1), X1(1), X2(1), X3(1), X4(1),
    X5(1)],'-bo')
    hold(app.UIAxes,'on')
    plot(app.UIAxes, zdoub, [-pin(1), -Y1(1), -Y2(1), -Y3(1), -Y4(1), -
    Y5(1)],':r')
    title(app.UIAxes,'Thin Lens (x,z) and (y,z)')
    xlabel(app.UIAxes,'z')
    ylabel(app.UIAxes,'x, -y')
    grid(app.UIAxes)
    legend(app.UIAxes,'x','-y','Location','SouthEast')
    hold(app.UIAxes,'off')
    %
end
```

A.8 Quad_Doublet_Thick_Thin

```matlab
function update(app)
    %
    % solve for and plot quadrupole doublet for 3 focal conditions
    %
    global Z Type
    %
    % Quadrupole Doublet - Thick and Thin Lens
    % pt -> pl (1), pl -> pt (2), Pt -> pt (3)
    %
    val = app.conditionDropDown.Value; % drop down condition
    tf = strcmpi(val,'->||');
    if tf == 1
        Type = 1;
    end
    tf = strcmpi(val,'||->') ;
    if tf == 1
        Type = 2;
    end
    tf = strcmpi(val,'->');
    if tf == 1
        Type = 3;
    end
    %
    % z Locations of Q1in, Q2in Constraint, both quad lengths fixed at
    5;% 3
    % 3 sliders - all quads of length 5 units
    Z(1) = app.Z1Slider.Value; Z(3) = app.Z3Slider.Value; Z(5) =
    app.Z5Slider.Value;
    Z(2) = Z(1) + 5; Z(4) = Z(3) + 5; % quads both of length 5
    %
    % starting values using thin lense
    % thin lens at CL of thick quadrupole
    %
    zq1 = Z(1) + (Z(2)-Z(1)) ./2; % quad center
    zq2 = Z(3) + (Z(4)-Z(3)) ./2;
    zcon = Z(5) - zq2;
    [ao(1), ao(2), xxx, yyy, zzz] = Doublet_Thin(app,zq1, zq2 - zq1,
    zcon);
    %
    plot(app.UIAxes,zzz,xxx,'-r',zzz,-yyy,'-g'); % plot thin lense
    results
    hold(app.UIAxes, 'on')
    app.fQ1thEditField.Value = ao(1);
    app.fQ2thEditField.Value = ao(2);
    %
    % Focal Lengths-Thin Lens, fx ,fy, ao(1), ao(2) )
    % use as starting vectors for thick lense
    %
    if Type == 1
        xo = [0.; 1.0]; % point
        yo = [0.; 1.0];
```

```matlab
      end
      if Type == 2
          xo = [1.0; 0.0]; % parallel
          yo = [1.0; 0.0];
      end
      if Type == 3
          xo = [0.; 1.0]; % point
          yo = [0.; 1.0];
      end
      %
      % fit to minimize matrix elements in x and y simultaneously
      %
      a = fminsearch(@app.Doublet_Fit,ao);
      %chi = Doublet_Fit(a);
      % Focal Lengths, fx, fy, a(1), a(2) )
      app.fQ1tkEditField.Value = a(1);
      app.fQ2tkEditField.Value = a(2);
      %
      % plot the fit
      %
      [zz,xx,yy] = Doublet_Plot(app,a(1),a(2),xo,yo);
      plot(app.UIAxes,zz,xx,':b',zz,-yy,':k')
      hold(app.UIAxes, 'on')
      plot(app.UIAxes,[zz(1) max(zz)], [0 0]) ; % x,y axis
      mxx = max(xxx);
      mny = min(-yy);
      plot(app.UIAxes,[Z(1), Z(1)], [mny-1 ,max(xx)+1])
      plot(app.UIAxes,[Z(2), Z(2)], [mny-1 ,max(xx)+1])
      plot(app.UIAxes,[Z(3), Z(3)], [mny-1 ,max(xx)+1])
      plot(app.UIAxes,[Z(4), Z(4)], [mny-1 ,max(xx)+1])
      xlabel(app.UIAxes,'z(m)')
      ylabel(app.UIAxes,'x,y - AU')
      title(app.UIAxes,'Quadrupole Doublet, Thin and Thick')
      legend(app.UIAxes,'(x,z)tn','(y,z)tn','(x,z)tk','(y,z)tk',
      'Location','Northwest')
      hold(app.UIAxes, 'off')

      %

  end



%---------------------------------------------------------------------
function[zz,xx,yy] = Doublet_Plot(app,f1,f2,xo,yo)
      %
      % plot thick quadruople doublet solutions
      % first quad x F - focal length f1, second x D - f2
      % z = 1,5 , 1 = quad enter, 2 = exit, 3 = enter 4 = exit, 5=focus
      % starting vectors xo, yo
      %
      global Z Type
      zz = linspace(0.,Z(5),50);
      %
```

```matlab
m1=[1.0, Z(1); 0.0, 1.0];
x1 = m1  xo;
y1 = m1  yo;
%
for i=1:length(zz)
    if zz(i) < Z(1);
        xx(i) = xo(1) + zz(i) .xo(2);
        yy(i) = yo(1) + zz(i) .yo(2);
    end
    if zz(i) > Z(1) && zz(i) < Z(2)
        b1 = 1.0 ./ (0.03 .f1 .(zz(i) - Z(1)));
        [k,f,phi,my1,mx1] = Quad_Matrix(app,b1,1.0,zz(i)-Z(1));
        x = mx1  x1;
        x2 = x;
        xx(i) = x(1);
        y = my1  y1;
        y2 = y;
        yy(i) = y(1);
    end
    if zz(i) > Z(2) && zz(i) < Z(3)
        xx(i) = x2(1) + (zz(i) - Z(2)) .x2(2);
        yy(i) = y2(1) + (zz(i) - Z(2)) .y2(2);
        m2 = [1.0, Z(3)-Z(2);0., 1.];
        x3 = m2  x2;
        y3 = m2  y2;
    end
    if zz(i) > Z(3) && zz(i) < Z(4)
        b2 = 1.0 ./ (0.03 .f2 .(zz(i) - Z(3)));
        [k,f,phi,mx2,my2] = Quad_Matrix(app,b2,1.0,zz(i)-Z(3));
        x = mx2  x3;
        x4 = x;
        xx(i) = x(1);
        y = my2  y3;
        y4 = y;
        yy(i) = y(1);
    end
    if zz(i) > Z(4) && zz(i) <= Z(5)
        xx(i) = x4(1) + (zz(i) - Z(4)) .x4(2);
        yy(i) = y4(1) + (zz(i) - Z(4)) .y4(2);
        m3 = [1.0, Z(5)-Z(4);0., 1.];
        x5 = m3  x4;
        y5 = m3  y4;
    end
  end

end


%---------------------------------------------------------------------------

function[matr] = Doublet_Fit(app,a)
    %
    % fit for quadruople doublet thick lens
```

```
%
global Z Type
%
% TYpe = 1 is point to par,2 = par to point,3 = point to point
%
% Z(1-5) are the 5 z boundaries
% a(1-2) are the 2 focal lengths - thin lense
%
m1 = [1.0, Z(1); 0.0, 1.0];
b1 = 1.0 ./ (0.03 .a(1) .(Z(2) - Z(1)));
[k,f,phi,mmy1,mmx1] = Quad_Matrix(app,b1,1.0,Z(2)-Z(1));
mx2 = mmx1  m1;
my2 = mmy1  m1;
m2 = [1.0, Z(3)-Z(2);0., 1.];
mx3 = m2  mx2;
my3 = m2  my2;
b2 = 1.0 ./ (0.03 .a(2) .(Z(4) - Z(3)));
[k,f,phi,mmx2,mmy2] = Quad_Matrix(app,b2,1.0,Z(4)-Z(3));
mx4 = mmx2  mx3;
my4 = mmy2  my3;
m3 = [1.0, Z(5)-Z(4);0., 1.];
mx5 = m3  mx4;
my5 = m3  my4;
%
if Type == 1
    matr = abs(mx5(2,2))+abs(my5(2,2));
end
if Type == 2
    matr = abs(mx5(1,1))+abs(my5(1,1));
end
if Type == 3
    matr = abs(mx5(1,2))+abs(my5(1,2));
end
%

end

%-----------------------------------------------------------------

function[k,f,phi,mx,my] = Quad_Matrix(app,bpr,p,l)
    %
    % finds quadrupole matrix elements - F(x) and D (y)
    % gradient bpr (kG/m), momentum p(GeV), quad length l(m)
    % find k in m^-2
    k = (0.03 .bpr) ./p;
    % find focal length f ~ 1/kl
    f = 1.0 ./(k .l);
    % find phase angle = √(k)l
    phi = √(k) .l;
    %
    mx(1,1) = cos(phi); % x is F
    mx(2,2) = cos(phi);
    mx(1,2) = sin(phi) ./√(k);
    mx(2,1) = -√(k) .sin(phi);
```

```matlab
    my(1,1) = cosh(phi); % y is D
    my(2,2) = cosh(phi);
    my(1,2) = sinh(phi) ./√(k);
    my(2,1) = √(k) .sinh(phi);
    %
end


%-------------------------------------------------------------------

function[f1, f2, xxx, yyy, zzz] = Doublet_Thin(app,L,l,Lo)
    %
    % thin lense approx for doublets - starting values for fits
    %
    global Z Type
    %
    % thin lense values for  D and F focal length
    % x is DF, y is FD,f(1) is first quad focal length, f(2) is second
    %
    fpttpl(1) = L .√(1 ./(L + 1)) ;
    fpttpl(2) = (1 .L) ./fpttpl(1);
    fpltpt(1) = √(1 .(1 + Lo));
    fpltpt(2) = (1 .Lo) ./fpltpt(1);
    c = L + 1 + Lo;
    fpttpt(1) = L .√((1 .(1+Lo)) ./((1+L) .c));
    fpttpt(2) = (1 .L .Lo) ./(c .fpttpt(1));
    % parallel to parallel f2-f1 = 1 or f1-f2 = 1 - degenerate and
    skipped
    %
    if Type == 1
        f1 = fpttpl(1);
        f2 = fpttpl(2);
        xo = [0.;1.];
        yo = xo;
    end
    if Type == 2
        f1 = fpltpt(1);
        f2 = fpltpt(2);
        xo = [1.;0.];
        yo = xo;
    end
    if Type == 3
        f1 = fpttpt(1);
        f2 = fpttpt(2);
        xo = [0;1];
        yo = xo;
    end
    %
    % x position and angle matrices - thin lense
    %
    m1 = [1., L; 0., 1.];
    m2 = [1., L;1.0 ./f1  ,1.0+L ./f1];
    m3 = [1.0 + 1 ./f1 , L + 1 + (L .1) ./f1;1.0 ./f1, 1.0 + L ./f1];
    m4(1,1) = m3(1,1);
    m4(1,2) = m3(1,2);
```

```
m4(2,1) = -1 ./(f1 .f2) + 1.0 ./f1 - 1.0 ./f2;
m4(2,2) = 1.0 + L ./f1 - (L + 1) ./f2 - (L .1) ./(f1 .f2);
m5(2,1) = m4(2,1);
m5(2,2) = m4(2,2);
m5(1,1) = 1.0 + 1 ./f1 - Lo ./f2 -(Lo .1) ./(f1 .f2) + Lo ./f1;
m5(1,2)= 1+L+Lo +(L .(Lo+1)) ./f1 - (1 .L .Lo) ./(f1 .f2)-(Lo
.(L + 1)) ./f2;
%
x1 = m1   xo;
x2 = m2   xo;
x3 = m3   xo;
x4 = m4   xo;
x5 = m5   xo;
%
% now the y vectors
%
f1 = -f1;
f2 = -f2;
m1 = [1., L; 0., 1.];
m2 = [1., L;1.0 ./f1  ,1.0+L ./f1];
m3 = [1.0 + 1 ./f1 , L + 1 + (L .1) ./f1;1.0 ./f1, 1.0 + L ./f1];
m4(1,1) = m3(1,1);
m4(1,2) = m3(1,2);
m4(2,1) = -1 ./(f1 .f2) + 1.0 ./f1 - 1.0 ./f2;
m4(2,2) = 1.0 + L ./f1 - (L + 1) ./f2 - (L .1) ./(f1 .f2);
m5(2,1) = m4(2,1);
m5(2,2) = m4(2,2);
m5(1,1) = 1.0 + 1 ./f1 - Lo ./f2 -(Lo .1) ./(f1 .f2) + Lo ./f1;
m5(1,2)= 1+L+Lo +(L .(Lo+1)) ./f1 - (1 .L .Lo) ./(f1 .f2)-(Lo
.(L + 1)) ./f2;
%
y1 = m1   xo;
y2 = m2   xo;
y3 = m3   xo;
y4 = m4   xo;
y5 = m5   xo;
%
% back to x,z plane
f1 = -f1;
f2 = -f2;
%
zzz = [0,L, L, L+1, L+1, L+1+Lo];
xxx = [xo(1), x1(1) , x2(1), x3(1), x4(1), x5(1)];
yyy = [yo(1), y1(1) , y2(1), y3(1), y4(1), y5(1)];
%
        end
```

Appendix B: Physics Constants

Electron rest mass	m_e	$9.109 \times 10^{-31}\,\mathrm{kg}$
Proton rest mass	M_p	$1.6726 \times 10^{-27}\,\mathrm{kg}$
Electronic charge	e	$1.6022 \times 10^{-19}\,\mathrm{C}$
Speed of light in free space	c	$2.9979 \times 10^{8}\,\mathrm{m\,s^{-1}}$
Permeability of free space	μ_0	$4\pi \times 10^{-7}\,\mathrm{H\,m^{-1}}$
Permittivity of free space	ϵ_0	$8.854 \times 10^{-12}\,\mathrm{F\,m^{-1}}$
Planck's constant	h	$6.626 \times 10^{-34}\,\mathrm{J\,s}$
Reduced Planck's constant	$\hbar = h/2\pi$	$1.0546 \times 10^{-34}\,\mathrm{J\,s}$
	$\hbar c$	$197.33\,\mathrm{MeV\,fm}$
Boltzmann's constant	k_B	$1.3807 \times 10^{-23}\,\mathrm{J\,K^{-1}}$

Quantity	Symbol, equation	Value	Uncertainty (ppb)
speed of light in vacuum	c	$299\,792\,458\,\text{m s}^{-1}$	exact*
Planck constant	h	$6.626\,070\,040(81) \times 10^{-34}\,\text{Js}$	12
Planck constant, reduced	$\hbar \equiv h/2\pi$	$1.054\,571\,800(13) \times 10^{-34}\,\text{Js}$	12
		$= 6.582\,119\,514(40) \times 10^{-22}\,\text{MeV s}$	6.1
electron charge magnitude	e	$1.602\,176\,6208(98) \times 10^{-19}\,\text{C}$	6.1,6.1
		$= 4.803\,204\,673(30) \times 10^{-10}\,\text{esu}$	
conversion constant	hc	$197.326\,9788(12)\,\text{MeV fm}$	6.1
conversion constant	$(hc)^2$	$0.389\,379\,3656(48)\,\text{GeV}^2\,\text{mbarn}$	12
electron mass	m_e	$0.510\,998\,9461(31)\,\text{MeV}/c^2 = 9.109\,383\,56(11) \times 10^{-31}\,\text{kg}$	6.2,12
proton mass	m_p	$938.272\,0813(58)\,\text{MeV}/c^2 = 1.672\,621\,898(21) \times 10^{-27}\,\text{kg}$	6.2,12
		$= 1.007\,276\,466\,879(91)\,\text{u} = 1836.152\,673\,89(17)m_e$	0.090, 0.095
deuteron mass	m_d	$1875.612\,928(12)\,\text{MeV}/c^2$	6.2
unified atomic mass unit (u)	(mass ^{12}C atom)/12 = $(1\,\text{g})/(N_A\,\text{mol})$	$931.494\,0954(57)\,\text{MeV}/c^2 = 1.660\,539\,040(20) \times 10^{-27}\,\text{kg}$	6.2, 12
permittivity of free space	$\epsilon_0 = 1/\mu_0 c^2$	$8.854\,187\,817\ldots \times 10^{-12}\,\text{F m}^{-1}$	exact
permeability of free space	μ_0	$4\pi \times 10^{-7}\,\text{N A}^{-2} = 12.566\,370\,614\ldots \times 10^{-7}\,\text{N A}^{-2}$	exact
fine-structure constant	$\alpha = e^2/4\pi\epsilon_0\hbar c$	$7.297\,352\,5664(17) \times 10^{-3} = 1/137.035\,999\,139(31)$†	0.23, 0.23
classical electron radius	$r_e = e^2/4\pi\epsilon_0 m_e c^2$	$2.817\,940\,3227(19) \times 10^{-15}\,\text{m}$	0.68
(e^- Compton wavelength)/2π	$\lambdabar_e = \hbar/m_e c = r_e\alpha^{-1}$	$3.861\,592\,6764(18) \times 10^{-13}\,\text{m}$	0.45
Bohr radius ($m_{nucleus} = \infty$)	$a_\infty = 4\pi\epsilon_0\hbar^2/m_e e^2 = r_e\alpha^{-2}$	$0.529\,177\,210\,67(12) \times 10^{-10}\,\text{m}$	0.23
wavelength of 1 eV/c particle	$hc/(1\,\text{eV})$	$1.239\,841\,9739(76) \times 10^{-6}\,\text{m}$	6.1
Rydberg energy	$hcR_\infty = m_e e^4/2(4\pi\epsilon_0)^2\hbar^2 = m_e c^2\alpha^2/2$	$13.605\,693\,009(84)\,\text{eV}$	6.1
Thomson cross section	$\sigma_T = 8\pi r_e^2/3$	$0.665\,245\,871\,58(91)\,\text{barn}$	1.4

Appendix C: Properties of Elements

Material	Z	A	$\langle Z/A \rangle$	Nucl. coll. length λ_T {g cm^{-2}}	Nucl. inter. length λ_I {g cm^{-2}}	Rad. len. X_0 {g cm^{-2}}	$dE/dx\vert_{min}$ {MeV g^{-1}cm^{-2}}	Density {g cm^{-3}} ({g ℓ^{-1}})
H$_2$	1	1.008(7)	0.99212	42.8	52.0	63.05	(4.103)	0.071(0.084)
D$_2$	1	2.0144101764(8)	0.49650	51.3	71.8	125.97	(2.053)	0.169(0.168)
He	2	4.002602(2)	0.49967	51.8	71.0	94.32	(1.937)	0.125(0.166)
Li	3	6.94(2)	0.43221	52.2	71.3	82.78	1.639	0.534
Be	4	9.0121831(5)	0.44384	55.3	77.8	65.19	1.595	1.848
C diamond	6	12.0107(8)	0.49955	59.2	85.8	42.70	1.725	3.520
C graphite	6	12.0107(8)	0.49955	59.2	85.8	42.70	1.742	2.210
N$_2$	7	14.007(2)	0.49976	61.1	89.7	37.99	(1.825)	0.807(1.165)
O$_2$	8	15.999(3)	0.50002	61.3	90.2	34.24	(1.801)	1.141(1.332)
F$_2$	9	18.99840316(6)	0.47372	65.0	97.4	32.93	(1.676)	1.507(1.580)
Ne	10	20.1797(6)	0.49555	65.7	99.0	28.93	(1.724)	1.204(0.839)
Al	13	26.9815385(7)	0.48181	69.7	107.2	24.01	1.615	2.699
Si	14	28.0855(3)	0.49848	70.2	108.4	21.82	1.664	2.329
Cl$_2$	17	35.453(2)	0.47951	73.8	115.7	19.28	(1.630)	1.574(2.980)
Ar	18	39.948(1)	0.45059	75.7	119.7	19.55	(1.519)	1.396(1.662)
Ti	22	47.867(1)	0.45961	78.8	126.2	16.16	1.477	4.540

(*Continued*)

Material	Z	A	$\langle Z/A \rangle$	Nucl. coll. length λ_T {g cm^{-2}}	Nucl. inter. length λ_I {g cm^{-2}}	Rad. len. X_0 {g cm^{-2}}	$dE/dx\vert_{\min}$ {MeV g^{-1}cm^{-2}}	Density {g cm^{-3}} ({g ℓ^{-1}})
Fe	26	55.845(2)	0.46557	81.7	132.1	13.84	1.451	7.874
Cu	29	63.546(3)	0.45636	84.2	137.3	12.86	1.403	8.960
Ge	32	72.630(1)	0.44053	86.9	143.0	12.25	1.370	5.323
Sn	50	118.710(7)	0.42119	98.2	166.7	8.82	1.263	7.310
Xe	54	131.293(6)	0.41129	100.8	172.1	8.48	(1.255)	2.953(5.483)
W	74	183.84(1)	0.40252	110.4	191.9	6.76	1.145	19.300
Pt	78	195.084(9)	0.39983	112.2	195.7	6.54	1.128	21.450
Au	79	196.966569(5)	0.40108	112.5	196.3	6.46	1.134	19.320
Pb	82	207.2(1)	0.39575	114.1	199.6	6.37	1.122	11.350
U	92	[238.02891(3)]	0.38651	118.6	209.0	6.00	1.081	18.950

Appendix D: Index of Refraction — Gases and Liquids

Medium	n
Gases at 0°C, 1 atm	
Air	1.000293
Carbon dioxide	1.00045
Hydrogen	1.000139
Oxygen	1.000271
Liquids at 20°C	
Benzene	1.501
Carbon disulfide	1.628
Carbon tetrachloride	1.461
Ethanol	1.361
Glycerine	1.473
Water, fresh	1.333

Appendix E: Table of Symbols

A	Angstrom, atomic weight, Area, Ampere
a	semi major axis of ellipse, accelerartion, beam pipe radius, nuclear size, PWC cathode radius
a_o	Bohr radius of hydrogen ground state
B	magnetic field
b	semi minor axis of ellipse, barn, impact parameter
C	capacitance, circumference of an accelerator
c	speed of light
C_L	Capacitance per unit length
D	diffusion coefficient, periodic dispersion, M(1,3) transfer matrix element
d	FODO distance between quadrupoles, distance from anode to cathode
$d\Omega$	differential solid angle
e	electron, electronic charge, exponential constant
E	electric field
eV	electron volt $= 1.6 \times 10^{-19}$ J
F	force
f	frequency, focal length
FFT	fast Fourier transform
FT	Fourier transform
G	gain of a circuit
g_m	transconductance of a transistor
h	Planck constant, r.f. harmonic number
$\hbar = h/2\pi,$	reduced Planck constant
I	electric current, ionization potential
J	angular momentum
j	current per area or length, current density, square root of -1
k	Boltzmann constant, wave number, quadrupole strength
K	kaon, quadrupole "kick"
$Linac$	Linear Accelerator
L	luminosity, length of an object, inductance
L_L	inductance per unit length

(*Continued*)

(Continued)

M	lattice transfer matrix, mass of a body
m	mass of a particle
m_e	electron mass
m_{pr}	projectile mass
n	number density, Bohr quantum number, index of refraction, neutron
N	total number of objects
N_A	Avogadro's number
p	particle momentum, proton
P	power
p_o	dipole moment
Q	betatron tune of an accelerator, charge of a system
q	particle electric charge in units of e
Q_s	synchrotron tune of an accelerator
q_s	source charge
r	coordinate radius
r_e	classical electron radius
r_M	Molière radius
$r.f.$	radiofrequency acceleration E field
R	radius of an object, electrical resistance
R_L	resistance per unit length
s	path length, square of CM energy of a system
S	Poynting vector, Slip Factor
T	absolute temperature, kinetic energy
t	time
TE	transverse electric field
u	energy density
U	total system energy, potential energy (NR)
U_I	energy deposited by ionization
v	velocity
V	volume, electrical voltage
v_d	drift velocity
v_T	thermal velocity
V_o	peak r.f. voltage
X_o	radiation length
Z	atomic number, electrical impedance
Z_b	beam impedance
z	longitudinal coordinate
α	fine structure constant, direction cosine, lattice parameter, BPM azimuthal angular size
β	velocity /c, lattice parameter
γ	photon, SR ratio of energy to mc^2, lattice parameter

(Continued)

γ_t	accelerator transition energy parameter
$\Delta p/p$	fractional shift in momentum mean
Δp_T	transverse momentum impulse
δ	skin depth, dp/p, Dirac function
ε	particle energy, emittance, permittivity
ε_c	critical energy
ε_o	vacuum permittivity, hydrogen ground state energy
ε_s	synchronous particle energy, multiple scattering energy parameter
η	dispersion parameter for a circular accelerator
θ	polar angle in spherical polar coordinates
θ_B	dipole "kick"
θ_R	Rutherford scattering angle
Λ	mean free path, Λ_I inelastic mean free path
λ	wavelength, deBroglie wavelength, Compton wavelength, charge/length
$\lambdabar = \lambda/2\pi$	reduced Compton wavelength
μ	mobility, betatron phase advance, muon, permeability
μ_o	vacuum permeability
ν	neutrino
π	pion
ρ	mass or charge density, radius of curvature, resistivity
σ	Gaussian standard deviation, cross section, conductivity, beam matrix $= \varepsilon\Sigma$
σ_R	Rutherford cross section
σ_T	Thomson cross section
Σ	beam matrix
τ	period of rotation, mean time between scatters, charge collection time, process time constant
φ	quadrupole phase factor
ϕ	azimuthal angle, quadrupole kick angle, r.f. phase
ϕ_B	bend angle in a dipole magnet
Φ	magnetic potential
ϕ_s	synchronous particle r.f. phase
ψ	general betatron phase
ω	circular frequency
Ω	ohm, solid angle
ω_o	accelerator rotation frequency
ω_p	plasma frequency
ω_{rf}	radio frequency circular frequency
$\sim$	approximately equal to
$\langle\rangle$	time average of a quantity

Appendix F: Acronyms

ALS — Advanced Light Source
ANL — Argonne National Laboratory
ATLAS — Experiment at the CERN LHC
BCT — Beam Current Transformer
BPM — Beam Position Monitor
CCD — Charge Coupled Device
CB — Conduction Band
CERN — European Center for Nuclear Research
CM — Center of Mass system
CMS — Experiment at the CERN LHC
cgs — units based on centimeters, grams and seconds
D0 — FNAL Tevatron experiment
DCA — Distance of Closest Approach
ENC — Equivalent Noise Charge
FFT — Fast Fourier Transform
FNAL — Fermi National Accelerator laboratory
FODO — Basic lattice repetitive element
FWHM — Full width at half maximum of a distribution
ID — Identification
IR — Infra-Red
IT — Information Technology
KEK — Largest Particle Physics Laboratory in Japan
LAr — Liquid Argon
LBNL — Lawrence Berkeley National Laboratory
LED — Light Emitting Diode
LEP — Large Electron-Positron, collider at CERN
LHC — Large Hadron Collider at CERN
mip — minimum ionizing particle
MKS — System of units used in this text, based on the meter, kilogram, and
  second
MR — Main Ring at FNAL
NR — non-relativistic
OTR — Optical Transition Radiation
PDE — Partial Differential Equation
PDG — Particle Data Group
PEP — Positron Electron Project at SLAC

(*Continued*)

(Continued)

PI — Particle Identification
PMT — Photomultiplier Tube
PWC — Proportional Wire Chamber
QE — Quantum Efficiency
QM — Quantum Mechanics
RICH — Ring Imaging Cerenkov Counter
r.m.s. — Root Mean Square of a distribution
SI — units using MKS and electromagnetic units using Coulombs, Volts, and Tesla
SiPM — Silicon Photomultiplier
SLAC — Stanford Linear Accelerator Center
SR — Special Relativity
STP — Standard Temperature and Pressure
TOF — Time of Flight
TR — Transition Radiation
UV — Ultra-Violet
UR — Ultra-Relativistic
VB — Valence Band
VDM — Van der Meer scan
WLS — Wavelength Shifting

References

In this text no attempt was made to supply an exhaustive set of references. Rather, the concept was to encourage the reader to go to a browser and look for specific details when additional information was desired. The search could be for more physics explanation or more detail about a particular device or class of devices. For many topics the "arXiv.org" is a very good place to start. Specifically, it is an open access site, as are now many other sites, which makes the topical content available to all. This site is updated daily and posts current research on many, many topics. Figure 1 shows the initial page at the site.

Figure 1: Initial page at the site where preprints in many areas of science are posted.

As mentioned in the text, the Particle Data Group is a great source for topics related to high energy physics. In particular their online site has a section on particle detectors. The references quoted there provide ways to explored further than the excellent section on particle detectors available at the PDG website itself. Specific physics topics are also discussed in the section on "Passage of Particles Through Matter" which also has references which are provided to encourage further study (Figure 2).

[1] T. Ferbel (ed.), *Experimental Techniques in High Energy Physics*, Addison-Wesley, Menlo Park, CA (1987).

[2] K. Kleinknecht, *Detectors for Particle Radiation*, Cambridge University Press, Cambridge (1998).

[3] G.F. Knoll, *Radiation Detection and Measurement*, 3nd edition, John Wiley & Sons, New York (1999).

[4] D.R. Green, *The Physics of Particle Detectors*, Cambridge Monographs on Particle Physics, Nuclear Physics and Cosmology, Cambridge University Press, Cambridge (2000).

[5] C. Leroy and P.-G. Rancoita, *Principles of Radiation Interaction in Matter and Detection*, World Scientific, Singapore (2004).

[6] C. Grupen, *Particle Detectors*, Cambridge Monographs on Particle Physics, Nuclear Physics and Cosmology, Cambridge University Press (2008).

Figure 2:   Reference from the PDG section on "Particle Detectors".

In the realm of particle accelerators there are annual schools in both the United States and Europe. A list of textbooks are shown in Figure 3 which appears on the USPAS web site, the US Particle Accelerator School.

Subject ⇕	Author ⇕	Book Title ⇕
Filter by Subject ▾	Filter by Author	Filter by Book Title
Accelerator Physics	S.Y. Lee	**Accelerator Physics - fourth edition** (World Scientific 2019)
Accelerator Physics	Valery Lebedev and Vladimir Shiltsev	**Accelerator Physics at the Tevatron Collider** (Springer 2014)
Accelerator Physics	Edmund Wilson	**An Introduction to Particle Accelerators** (Oxford University Press 2001)
Accelerator Physics	Donald A. Edwards and Michael J. Syphers	**An Introduction to the Physics of High Energy Accelerators** (Wiley & Sons Publishers 1993)
Accelerator Physics	Mario Conte and William M. MacKay	**An Introduction to the Physics of Particle Accelerators - second edition** (World Scientific 2008)
Accelerator Physics	Ken Takayama and Richard J. Briggs eds.	**Induction Accelerators** (Springer 2010)
Accelerator Physics	Stephen Peggs and Todd Satogata	**Introduction to Accelerator Dynamics - first edition** (Cambridge University Press 2017)

Figure 3:   Short list of textbooks from the USPAS web site for further reading on accelerator topics.

A site for the European accelerator school also has links to very useful references to many topics of interest. A partial list appears in

Figure 4. These many resources afford the user a very rapid feedback when searching for more information on any specific topic.

All	Introductory	Advanced	Specialised	Basic

Year	School title	City	Country
2019	CAS@ESI: Basics of Accelerator Physics and Technology	Archamps	France
2019	Introduction to Accelerator Physics	Vysoke-Tatry	Slovakia
2019	Advanced Accelerator Physics	Slangerup	Denmark
2019	High Gradient Wakefield Accelerators	Sesimbra	Portugal
2018	Numerical Methods for Analysis, Design and Modelling of Particle Accelerators	Thessaloniki	Greece
2018	Introduction to Accelerator Physics	Constanta	Romania
2018	CAS@ESI: Basics of Accelerator Physics and Technology	Archamps	France
2018	Beam Instrumentation	Tuusula	Finland

Figure 4: Website list of resources which can be downloaded for specific topics of interest in accelerator physics. Many more resources are available.

Index

acceleration, 16, 22, 23, 72–74, 81, 84, 90, 127, 128, 138, 139, 143, 190, 199, 201, 204, 235

ALICE, 47, 60

amplitude, 12, 13, 45, 55, 72–74, 77, 107, 122, 140

anode, 41, 93, 94, 96, 100, 102, 104, 105, 234

aperture, 123, 124, 126, 127, 138, 174, 175

avalanche, 43

Avogadro number, 13, 235

backscatter, 69, 71, 207–209

beam dump, 81

beam matrix, 125, 126, 165–167, 180, 182–185, 236

beam impedance, 235

bend angle, 18, 115, 116, 132, 135, 136, 138, 182, 205, 236

Bessel, 4, 157, 200

binding energy, 7, 9, 11, 12, 37, 39

BPM, 130, 152, 155–159, 161–170, 172–174, 176, 181–195, 197, 211, 235, 237

Bragg curve, 61

bucket, 139, 140, 143, 190, 194

bump magnets, 186

bunch, 16, 139, 143–145, 151, 152, 158, 159, 171, 172, 186, 187, 191, 194, 196–198, 204, 205

bunch length, 143–145, 154, 158, 159, 179, 180, 194–198, 204, 208

calibrate, 81

calibration, 44

cascade, 84, 85, 87, 96

cathode, 39, 41, 94, 95, 97, 99–102, 105–107, 111, 234

cathode strip, 99, 106, 107, 111

center of momentum, 33, 185

chromaticity, 24, 132, 161, 162, 180, 190

classical electron radius, 8, 230, 235

coasting beam, 190, 194

coaxial cable, 112–114, 174, 179, 194, 195

coherence, 72, 75, 79

coherent, v, vii, 27, 55, 73, 77, 204

collider, v, ix, 20, 66, 84, 154, 184–186, 203, 237

collimator, 24, 45, 174, 175

Compton radius, 13

Compton scattering, 8, 68, 69, 71–73, 205

conduction band, 101, 237

constant of motion, 142

cooling, 84

critical energy, 78–81, 84, 85, 205, 236

crossed, 22–24, 31

current source, 108, 109, 151, 153